MY LIFE IN CATTLE RANCHING

MY LIFE IN CATTLE RANCHING

CHARLES E. BACKUS

Clear Creek Publishing, Tempe, Arizona

My Life in Cattle Ranching

Copyright © 2025 by Charles and Judith Backus,
and Clear Creek Publishing.

Chuck and Judy Backus
Tempe, Arizona

All Rights Reserved.
First Printing: January 2026

All rights reserved. No part of this book may be reproduced or transmitted in any form or by any means, electronic or mechanical, including photocopying, recording, or by any information storage and retrieval system without written permission from the authors, except for the inclusion of brief quotations in a review.

Clear Creek Publishing
P.O. Box 24666
Tempe, Arizona 85285 U.S.A.
www.ClearCreekPublishing.com

ISBN:
978-1-884224-19-5 (hardcover)

Library of Congress Control Number: 2025903110

Printed in the United States of America

Cover and Interior design: 1106 Design, Phoenix, Arizona

Publisher's Cataloging-in-Publication Data

Names: Backus, Charles E.
Title: My life in cattle ranching / Charles E. Backus.
Description: Tempe, AZ : Clear Creek Publishing, 2025. | Includes color photos and illustrations.
Identifiers: LCCN 2025903110 | ISBN 9781884224195 (hardcover)
Subjects: LCSH: Quarter Circle U Ranch (Ariz.). | Beef cattle – Breeding. | Cattle trade. | Animal culture. | Ranch life – Arizona – Biography. | BISAC: BIOGRAPHY & AUTOBIOGRAPHY / Business. | BUSINESS & ECONOMICS / Industries / Agribusiness. | SCIENCE / Life Sciences / Zoology / Mammals.
Classification: LCC HD9433.A2 B33 2025 | DDC 338.1762--dc23
LC record available at https://lccn.loc.gov/2025903110

Front Cover Photo: During the 6-week backgrounding period of the calves, Chuck walks through the calves to get them more comfortable with people. Here he is lecturing them on how he expects them to perform in the feedlot or to perform as replacement cows. Photo by Dean Harris, 2015.

Back Cover Photo: The Backus family, *right to left*: Chuck Backus with wife Judy, son Tony with wife Blanca, daughter Beth, and daughter Amy with husband Mike Doyle. They all ride and help with the ranch. Photo by Pam Lannan, 2013.

From left: Judy, Tony, Amy, Chuck, and Beth.

To my wife, Judith Ann Clouston Backus (Judy), and our three children, David Anthony Backus (Tony), Elizabeth Ann Backus Roth (Beth), and Amy Jo Backus Doyle (Amy), who have made my life enjoyable, meaningful, and productive. All of them worked with me during those many years at the ranch. I could not have operated it without them.

Contents

Dedication v
Preface xiii
Introduction xv

CHAPTER 1 Cattle Ranching in the Americas and Arizona 1

CHAPTER 2 My Life Before Buying the Quarter Circle U Ranch . . . 5
    Backus and Strader Farms, Nicholas County, West Virginia 5
    Theron Cooperrider Dairy Farm, Croton, Ohio 6
    Grad School, Westinghouse, and ASU 7

CHAPTER 3 Early History of the Quarter Circle U Ranch 11
    Matt Cavaness Establishes the Ranch 11
    George Marlow and Alfred Charlebois 14
    Jim Bark and Frank Criswell 14
    Barkley Era 16
    Private Land Added to the Barkley Ranch 20
    Preparation for the Sale and Division of the Barkley Ranch . . . 21
    Barkley Ranch Is Divided. John Cox and Joe Lamb 22
    Joe Lamb and Guy Hill 23

CHAPTER 4 Backus Purchase and Early Operation 25
    Purchase of Guy Hill's Foreclosed Ranch 25
    The USFS Superstition Allotment 26

    The Trial in Florence, Arizona 27
    Purchase of Headquarters 10 Acres 28
    Purchase of Remaining Private Ranch Land 29
    Early Evaluation and the NRCS 31
    Transect Monitoring 33
    Improvements at the Headquarters 37
    Fire at the Stone Barn 40
    Evolution of Branding Methods 42
    Building the Tack Room 44

CHAPTER 5 The Beef Co-op and My Plan to Expand 47
    The Arizona Natural Beef Co-op 47
    My Bad Decision 48

CHAPTER 6 My Ranch Managers 51
    Chuck Sanders 51
    Henry Jones 52
    Arkie Johnson 54
    Herb Herbert 56
    Howard Horinek 56
    Dean Harris 59
    Jordan Selchow 60

CHAPTER 7 Ranch-Related Events .63
    A&M Universities and the Creation of ASU 63
    Equipment Purchased from the Closure of the ASU Farm 64
    More Construction Material Acquired 66
    Abundance of Wildlife on the Ranch 67
    Events for Ranch Family and Guests 73

CHAPTER 8 Range and Water Improvements
    at the Quarter Circle U Ranch 75

Contents

CHAPTER 9 Superstition and Northern Ranches, Grazing Leases, and Allotments 105

Superstition Quarter Circle U Ranch 105
Northern Ranch 106
Improvements in the Pastures 107
Headquarters for the Northern Ranch 111

CHAPTER 10 Headquarters Handling Facilities 117

Headquarters Corral Design 117
The Installation 119
Operation of the Processing Facilities 122

CHAPTER 11 Cattle Operations Schedule 137

October–Processing Cattle from Summer Pasture 138
October–Calf Processing 140
October–Adult Cow Processing 142
October–Bulls Put in Separate Pasture 144
November and December–Calves Sent to the Feedyard . . 144
December to February–Perennial and Annual Grass Natural Feed 145
December–Tagging the Calves 146
February–Bull Testing 148
February and March–Artificially Inseminating Cows and Heifers . . 152
April–Bull Selection 153

CHAPTER 12 Calf Processing in the Spring 157

Freeze Branding 157
Electronic Identification (EID) Tags 161
Calf Castration 162
Heifer Calf DNA Samples 162
Calf Vaccinations 164
Calf Dehorning 164
Weaning 164

CHAPTER 13 Cattle Operations: Artificial Insemination **165**

    The Case for Using Artificial Insemination **165**

    Artificial Insemination: Fixed-Time AI **167**

    Scheduling the AI Protocol **172**

    AI Technicians **173**

    First Evaluation of AI and New Goal Set **173**

    Retaining Yearlings for Future Herd Cows **174**

    Live Bull Selection **175**

    AI Bull Semen Selection **176**

    Documentation **178**

CHAPTER 14 Cattle Improvements: General Considerations . . **181**

    Need to Learn About Tools **182**

    Beef Grading **182**

    Records and Measurements **183**

    Development of EPDs **183**

    Setting, Measuring, and Commitment to Goals **184**

    Bull Genetics **186**

    Genome Development and Enhanced EPDs **186**

    Cattle Traits and Heritability **187**

    Feedlots Want Healthy Calves that Bring Premiums **188**

    Packing Plants **190**

    Summary **193**

CHAPTER 15 Herd Improvement: Bull Selection and Other Methods **195**

    Selection Indexes **195**

    Prioritize Several Traits **196**

    Physical Environment **196**

    Numbers Based Decisions **197**

Setting First Goal and Secondary Traits 197
EPD Comparison Within the Breed 198
DNA Testing for Replacement Heifer Selection 198
Herd Improvement by Culling Mother Cows 200
Herd Improvement by Increasing the Quality
  of the Beef Produced 200
Herd Improvement by Increasing Feed Conversion Efficiency . . 202
Herd Improvement by Using Artificial Insemination (AI) . . . 203
Herd Improvement by Herd Management 204
Other Considerations for Herd Improvement 204

CHAPTER 16 Herd Improvement: My Personal Experience . . . 207
Ranch Goal 208
Changes Needed to Meet My Goal 208
Adopted Changes 210
My Mentors 212
Selecting a Feedyard 213
Herd Goal for Premiums Achieved 214

CHAPTER 17 Solar Installations at the Ranches 219

CHAPTER 18 Continuing Education for Commercial Ranchers . . 227

CHAPTER 19 The Transfer of Ranch Ownership 233

APPENDIX A How to Implement Your Herd Goal 235

APPENDIX B *Expected Progeny Differences* 239

APPENDIX C *Interpretation and Use of Expected
  Progeny Differences* 247

About the Author 259

Preface

In 2024, I wrote *My Multi-Faceted Life*, which was a 200-page hardcover book with color photographs. It was written for my descendants and close friends, and it told the story of my entire life. The longest chapter was about my experiences in cattle ranching; however, there wasn't room to include many of the improvements to my ranch and the cattle herd.

Chuck and Judy Backus on the back patio of the Quarter Circle U ranch house. Photo by Steve Suther, Certified Angus Beef (CAB), April 2013.

I realized at the time that I needed to write a new book that covered only the ranching aspect of my life. This book, *My Life in Cattle Ranching*, includes the improvements to the ranch and herd, more photographs, and the events throughout my life that prepared me for life as a rancher.

The purpose of this book is to describe my research-driven science for cattle genetics and about fine-tuning bull selection. Most books about cattle are textbooks intended to be used in classes for students in a classroom setting. This book may be of interest to commercial cattle ranchers, classrooms focused on ranching, or for supplemental reading. It also might be ideal for people considering buying a cattle ranch, for those who just bought one, for those interested in the application of cattle genetics for improving their herd, and for learning about other changes in the cattle industry.

Judy and Chuck Backus inside the Quarter Circle U Ranch bunkhouse.

I hope this culmination of my learning and implementation of progressive cattle ranching will be helpful to others.

Introduction

My wife Judy and I have owned the Quarter Circle U Ranch for more than 45 years, during which time we have made many physical changes and improvements. We bought this old historic ranch, which was established in 1876, while I remained a full-time professor of engineering at Arizona State University in Tempe, Arizona.

I operated the ranch by hiring a full-time ranch manager and gave him instructions on what actions should be done during the week, with me being there on weekends and holidays to direct the ranch operations. I continued with this schedule for 27 years until my retirement at age 66 from ASU in 2004.

After retirement, I became a full-time rancher, and I had time to do a serious study on where the cattle industry was going and into bovine genetics. I then decided what improvements could be made at the Quarter Circle U Ranch for us to become a leader in the cattle industry in Arizona.

The cattle industry has changed rapidly in the last 20 or more years, primarily because of the advancements in the understanding of cattle genetics, the development and use of expected progeny differences (EPDs), and more complete measurements of physical parameters. Even during the twentieth century, the purebred/registered cattle people selected their purebred bulls on their appearance. Very few measurements were recorded by purebred producers when they listed the animals for sale at registered bull sales.

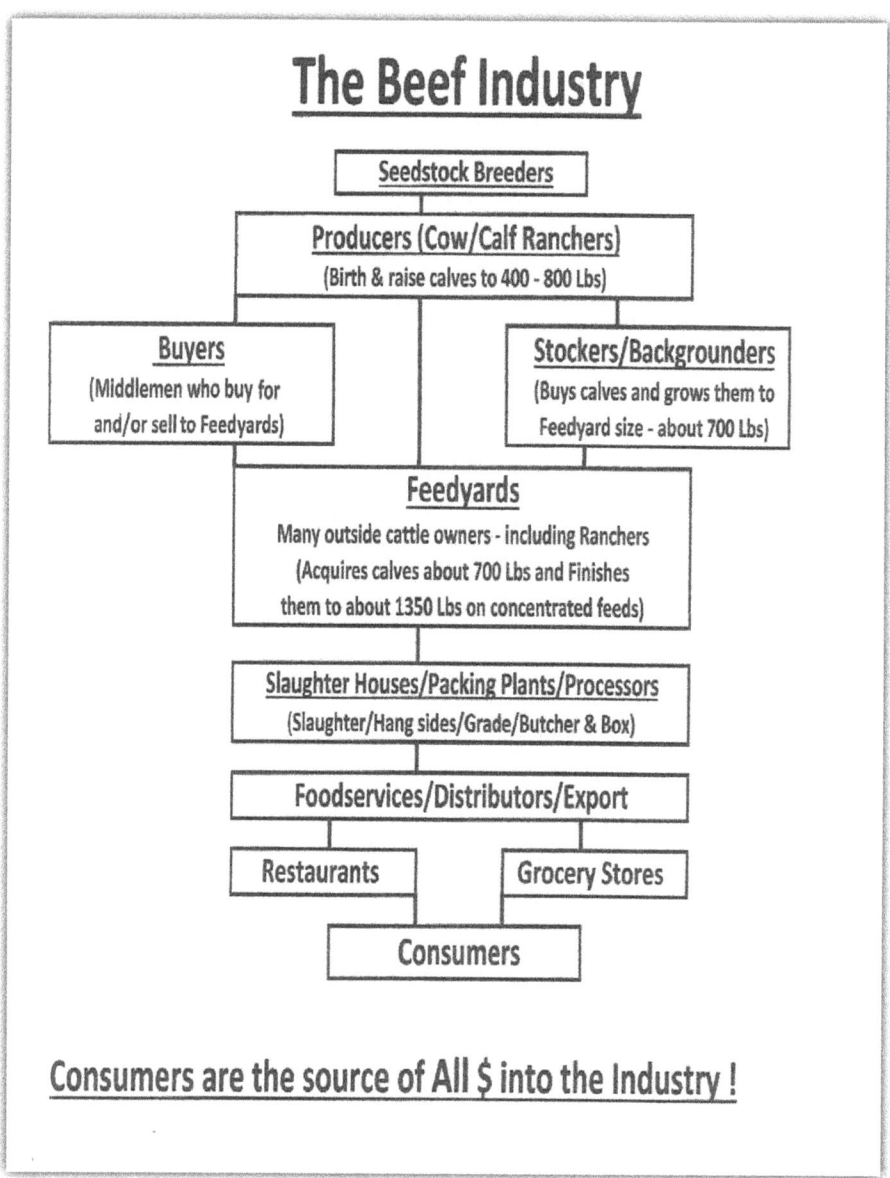

This flowchart shows why ranchers need to focus their herd goals to meet the needs of the consumer. The cattle processing starts at the top of the chart with the seedstock breeders and moves toward the consumer at the bottom. However, the money from the consumer flows from the bottom of the chart and moves up toward the top. Chart by Chuck Backus.

Today, bull sales are dominated by measurements made on individual bulls, including DNA results and predictions for the characteristics of their calves (EPDs, DNA, and enhanced EPDs).

I concluded that the application of recent research on bovine genetics and breeding could even affect the way that our commercial cattle should or could be raised. I decided to incorporate these progressive breeding techniques into my cattle operation to improve the quality of my herd's productivity and thus directly compete with the Midwestern cattle farmers.

Commercial cattle ranchers like me do not have to replicate university science, but we need to use the tools being developed; otherwise, we may find ourselves uncompetitive.

The selection of bulls and the selection of calves for the replacement of cows in the herd are the two most important parameters that I and other commercial ranchers can control that will affect the characteristics of the cattle and the resulting beef.

An individual cow can give birth to 5 or 8 calves in her lifetime and influence their genetic characteristics. A herd bull can influence more calves than that per year, with a total of 30 to 100 calves during his lifetime in a herd. An AI bull can influence the genetic characteristics of several hundred calves in a herd. Thus, the primary way to improve a herd is with an emphasis on live bull selection or semen selection of a bull being used for artificial insemination.

Chapter 1

CATTLE RANCHING IN THE AMERICAS AND ARIZONA

The technique of cattle ranching, as opposed to raising cattle, originated in the 1200s in southern Spain. This region had large expanses of grasslands that were unsuitable for farming.

Ranching is defined as allowing cattle of multiple owners to graze on unrestricted range, tending to them by horseback and having roundups, and then having rules and organizations to establish ownership and to resolve conflicts. This evolved in the US into mostly having ranches owned by only one rancher.

On Christopher Columbus's second voyage to the Americas in 1493, he brought Spanish cattle to the Caribbean Islands of the Americas. Cattle were introduced into mainland New Spain (present-day Mexico) in 1521. Because cattle are so adaptable and can digest any type of vegetation, they prospered very well. Since no fences existed in all of the Americas, these unrestricted cattle ranching techniques resulted in many stray cattle that eventually developed into wild herds. By 1540, some of these wild cattle herds were reported as far north as the Rio Grande Valley of what is now the United States. The Coronado Expedition that entered Northern New Spain (present-day Arizona) in 1540 looking for cities of gold included 150 head of cattle to provide fresh food for their army.

However, large cattle herds were not introduced into present-day Arizona until the 1600s, when Spanish missionaries used them as a tool to try to domesticate the Indians. For example, Father Kino brought a herd of seven hundred cattle to San Xavier del Bac. Beef became a staple food for both the missionaries and also for the raiding Apache Indian tribes. The Apaches considered the missions more like shopping centers. The extent of cattle ranching in present-day Arizona in the 1700s might be inferred from the report from the Commandant of the Tubac Presidio in 1770, when he reported that recent Apache campaigns had resulted in the death of 115 hostiles and the recovery of 2500 stolen cattle.

During the period from 1830 to 1850, the Apache tribes eliminated all of the cattle herds in present-day Arizona by selectively eating only the calves and cows. They did not care for the taste of bull meat. In the 1850s, the Mormon Battalion, the Army of the West, and some government survey parties reported being harassed and damaged by herds of wild bulls. After the 1850s, the only cattle in the Arizona Territory were those that were driven through the territory on their way from Texas to the California gold fields.

After the American Civil War was over, the U.S. Army returned to the Arizona Territory to provide some protection; thus, miners from California and settlers from the East started moving into the Arizona Territory. This growth created a demand for beef, and a few permanent cattle ranches were established. One of those was the Quarter Circle U Ranch on the southern edge of the Superstition Mountains.

Cattle are ruminant animals, which is their key advantage. Since their rumen can digest almost any plant-based material, cattle can survive in almost any environment that grows plants. Their multiple stomachs allow them to convert inferior proteins, such as plants, into the superior proteins of milk and meat.

Their ability to survive on any kind of plant feed source allows cows to live in semi-arid areas, in the valleys and slopes of high mountain areas, as well as in southern tropical areas. Most of the landscape in the Western United States is not suitable for cultivation or high-intensity grazing but is suitable for raising high-quality beef on sparse feed. Thus, most of the large ranches in the United States are in the Western states. This book is applicable to all areas that raise cattle.

Chapter 2

My Life Before Buying the Quarter Circle U Ranch

Backus and Strader Farms, Nicholas County, West Virginia

I have descended from many generations of farmers. My father, grandfather, and great-grandfather were born on a farm in Nicholas County, West Virginia. My family lived close to both of my grandparents' farms, and that is where I spent most of my summers and school holidays. My grandparents Strader's farm was so much larger and more modern than my grandparents Backus's farm.

During the summer, when I was off from school, I would often spend two or three weeks at a time at my Strader and Backus' grandparents' farms, between my parents' visits. This was especially true during the haying season, which lasted most of the summer. They never had any tractors, so everything was done by horses. All they ever had, while I was growing up there, were workhorses—no horses for riding. Workhorses pulled the corn and hay wagons, sleds, mowers, and rakes.

They did not let me mow the open grazing pastures on the steeper slopes, thinking it was dangerous for a young boy. But driving a horse-drawn rake to rake cut hay into windrows or to drive horses pulling a sled along a hay row was okay. When I was raking hay into windrows,

I would have to jump up with both feet on the handle to get enough weight on the handle to lift the rake up when coming up to the end of a windrow—just to get the rake to raise—while still driving the horse.

Using a pitchfork to load hay onto a sled or from the sled into the barn was also okay for young boy. When we were hauling hay to a haystack, it was us small ones that got assigned to stomp the hay on the haystack and shape it into a rounded, peaked stack.

All of us children—me, my sister, and my many cousins—went to the fields and worked during corn hoeing time and for hay and oats raking. The job I hated the most was hoeing field corn. Grampa would drive a horse-drawn plow between corn rows, but we had to hand-hoe between hills of corn. Harvesting the field corn was a lot more fun than hoeing it. We would rip the ears off the corn stalk, and it was easy to peel the dried coverings from the ear and then throw the ear of corn into large piles. We would later come by with a sled, load the ears into the sled, and haul the corn to the barn for storage and winter feeding.

We picked apples and berries and peeled apples at night until bedtime. I carried water from the spring, and adults and children alike washed clothes on washboards with rainwater from barrels. In summer, we children liked to square dance with the adults and neighbors on the lawn. We also picked beans, put them in sacks, and rolled them down the hill—with us inside the sacks.

Theron Cooperrider Dairy Farm, Croton, Ohio

We moved to Croton in rural Ohio for my last two years of high school. Theron Cooperrider, a member of the church greeting committee at Croton, offered me a job to work on his dairy farm for the summer. I worked through that summer and on Saturdays and holidays through the next school year.

When my family was moving to Haydenville (next door to my future wife, Judy), Mr. Cooperrider told me that if I wanted to finish my senior year in high school at Croton, I could live with them on the farm in exchange for work for the summer and for the entire school year until spring graduation. That turned out to be a bad decision for my college preparation since there were 15 people in my class, and very few courses were taught for college-bound students.

Mr. Cooperrider had tractors and other machinery. It was not like my grandfathers' farms, where everything was drawn by horses. They had balers pulled by tractors, with wagons behind the balers to stack the bales on. They had automatic milkers that we connected to the cows. We emptied the milk into cooled containers for temporary storage. The milk trucks came by each week to load the cans of cold storage containers.

It was a complete dairy farm, meaning that they grew all the feed for their 125 dairy cows. Thus, I experienced the entire farm operation for crops—hay, grain, and corn—and equipment, as well as working with cows. All the individual cows had names and lifetime records. I did not learn all of the cows' names, but Mr. Cooperrider knew each one at first sight.

I worked for Mr. Cooperrider all summer, except Sundays, and during the school year on weekends and days off. The two years of helping on the dairy farm were valuable and satisfied my farming instincts.

Grad School, Westinghouse, and ASU

During my sophomore year at Ohio University in Athens, Ohio, the Russians launched the first space satellite using nuclear isotopes as a power source, and it stimulated my interest in that science.

I graduated from Ohio University in Athens, Ohio, in 1959 with an engineering degree and then enrolled in the University of Arizona (UA) graduate school in nuclear engineering to study power systems with nuclear power sources. I worked half-time for the UA and taught laboratory classes in nuclear science. I also operated the nuclear reactor for various college classes. I did graduate studies at the UA for five years and had summer jobs at US government national labs around the western states. I spent most of my free time hiking and climbing—mostly in Arizona, Idaho, Montana, California, New Mexico, and Mexico.

Upon completion of my PhD at the UA in nuclear engineering, I worked at the Westinghouse Astro-Nuclear Laboratory in Pittsburgh, Pennsylvania. While I worked there, I was on assignment for one year at Aerojet General in Sacramento, California. I spent all of those weekends hiking and mountain climbing in the northern Sierra Mountains.

During my times of hiking in the West, I was stimulated to think that I would like to own a cattle ranch in the West so I could continue spending a lot of time outdoors. When I left Westinghouse in 1968 to become a professor of engineering at Arizona State University (ASU) in Tempe, Arizona, owning a cattle ranch was still on my mind.

The ASU College of Engineering and Applied Sciences included a School of Agriculture that taught courses in Animal Science and Range Management. After establishing myself in my engineering professor's job for about two years, I started auditing Ag classes in Animal Science and Range Management—one course each semester. I did not formally audit the courses but just asked the professor if I could sit in the back of the class and that I would not ask questions, just absorb the information. I read the textbooks outside of class. I must have sat in and read textbooks for about seven or eight courses on ranching, and that better prepared me to become a rancher.

I also learned that the dean of the engineering college, Dr. Lee Thompson, was not only raised on an Arizona ranch but had always owned a ranch in Arizona while being the founding dean of the engineering college. I thus waited for the opportunity, some years later, to discuss my interest in ranching and to seek his advice. He suggested two things: (1) that I buy a truck, which I did, and (2) that I make an appointment to speak with his son-in-law, Dwayne Webb.

Engineering Dean Lee Thompson (right) congratulating Chuck (left) and Dr. Charles D. Hoyt (center) after they received the two university-wide awards given for the year 1976. Dr. Hoyt was selected as the best teacher of the year at ASU, and Chuck was selected as the outstanding professor "for Academic Contribution Beyond the Classroom" (meaning for research). Dr. Thompson was also a lifelong cattle rancher and later became a father figure to Chuck.

Dwayne was a loan officer in the Production Credit Association (PCA), which is now called Farm Credit. This organization loaned money to farmers and ranchers, and thus Dwayne knew all about the availability, the operation, and the costs of buying and running ranches.

I met with Dwayne and described my interest in becoming a rancher. However, I did give him some constraints. I had a reasonable net worth with ownership of the mobile home park where we lived, but I had limited cash available. I also needed a ranch relatively close to Phoenix because I planned to keep my job at ASU. The good news was that I had a high-paying, extremely stable job and didn't need to have a ranch that financially supported itself.

My story about buying the Quarter Circle U Ranch continues in Chapter 4, which explains how Dwayne contacted me about the upcoming foreclosure of that ranch in 1977.

An early morning view to the south of the eastern half of the Quarter Circle U Ranch headquarters. The hay barn is in the foreground with the old stone barn and windmill farther back. The tack room is on the right edge of the photo. The Lady's Nose Mountain is south of the headquarters. The mountain is named after the rock formation on the horizon that looks like the face of a person with a large nose looking upward. Photo by Steve Suther, Certified Angus Beef (CAB), April 2013.

Chapter 3

Early History of the Quarter Circle U Ranch

Matt Cavaness Establishes the Ranch

The Silver King Mine opened near Pinal City, Arizona (near the present-day town of Superior, Arizona), in 1875. Matt Cavaness was the teamster hired to haul the silver ore from Pinal City to Yuma, Arizona. In Yuma, it was transferred to a steamboat, which took the ore to a smelter in San Francisco, California. Cavaness used 20 mule or 16 oxen team wagons to haul the heavy ore over very primitive roads. Cavaness became aware of the increasing demand for beef to serve both the Silver King miners and the growing town of Phoenix. When an "old Indian" told Matt about a reliable water source on the southern edge of the Superstition Mountains, which was located between those two beef demand markets, he decided to take advantage of it. He bought 1,000 head of cattle to start a ranch at that reliable water site. That site is where the Quarter Circle U Ranch house is located.

On a return trip from Yuma, Matt brought back sawed lumber and built a house and corrals at the water site. Houses outside of the cities that were made of sawed lumber were extremely unusual, so this ranch house became known to everyone as the *plank house*. The house (see figures 3.1 and 3.2) was one room with a front porch and an attached kitchen and bedroom. The house was built in 1876 and essentially remains in use today with very few modifications. Matt moved his wife and family into this ranch house and then opened a butcher shop in Pinal City as an outlet for his beef. Since Matt continued to haul ore for the Silver King, he was on the road most of the time. Alice Cavaness, living in the ranch house, also raised some milk cows at the ranch and began selling milk to people in the surrounding area. While Matt was away hauling ore and running a butcher shop in Pinal City, the running of the ranch was left to Alice, their three young sons, and a few hired hands.

Figure 3.1. This is the plank house *that Matt Cavaness built in 1876. It received its nickname because it was made from lumber that Matt hauled from Yuma. It is now the Quarter Circle U Ranch headquarters. Photo taken about 1931 with a view to the north to the Dacite Cliffs. Photo courtesy of Gregory E. Davis. The original photo was from the collection of Odd S. Halseth. This may be a photo taken by Ed Newcomer during the search for Dr. Adolph Ruth.*

In those days, the ranch boundaries were not defined, and ranchers typically ran cattle in the country surrounding their ranch houses or headquarters. The ranches often had outlying shacks that would serve as overnight sites for cowboys to stay in while gathering or working cattle in that area. The Cavaness Ranch essentially extended south to what is now US 60, north to what is now Canyon Lake on the Salt River, west to what is now called Apache Junction, and east to Pinal City near present-day Superior. His range was about 150 square miles. At that time, all land was considered open range with no ownership except for the possessory rights claimed by people who used the land for their cattle herds. Even the land at the Quarter Circle U Ranch headquarters did not become private land until 1947.

Figure 3.2. Mathew "Matt" Cavaness. Matt established the Quarter Circle U Ranch in 1876. He was a freighter and entrepreneur of many businesses in the Pinal City area and later in Tonto Basin. Photo courtesy of Jack San Felice and Gregory Davis.

The loneliness and hardship of Alice in running the ranch eventually led to her attraction to another man and to the divorce of Alice and Matt on October 21, 1881. Alice sold the ranch to merchant Leo Goldman of Phoenix. She married Randolph Perdue on November 22, 1881.

Matt Cavaness died in 1929 at the Sawtelle Veterans Home in California. In his later years, he wrote his life story in a longhand manuscript titled *Memoirs of Matt Cavaness*, which later was typed by Joseph Miller. I have a copy of these notes.

George Marlow and Alfred Charlebois

George Marlow and Alfred Charlebois purchased the Cavaness Ranch from Leo Goldman on July 22, 1882, for $200. They were the proprietors of the Pinal Meat Market and the Palace Hotel in the town of Pinal. In March 1883, Marlow purchased the full interest of the Cavaness Ranch from Charlebois for $200. During the late 1880s, this ranch was known as the ML Ranch for the ML brand name but often referred to in various maps and articles of the time as just the Marlow Ranch.

Jim Bark and Frank Criswell

In early 1891, Jim Bark bought the ranch from the Marlow estate, and he is thought to be the one responsible for bringing the Quarter Circle U branding iron to this ranch. For more than the last 134 years, this ranch has gone by the Quarter Circle U Ranch name. The Quarter Circle U brand left the ranch when the Barkley family sold the ranch, the brand, and

Alfred Charlebois and George Marlow were partners in the former Cavaness Ranch, which became the Quarter Circle U Ranch. From July 22, 1882, to March 30, 1883. Charlebois and Marlow supplied their Pinal Meat Market with cattle from the Cavaness Ranch. They were also partners in the Palace Hotel and other businesses in the Pinal City area. Pinal Drill, *Nov. 20, 1885. Courtesy of Jack San Felice.*

the cattle in 1967. It is interesting that the name, Quarter Circle U Ranch, remained with the headquarters on private land, although the brand was now used on John Cox's Forest Service grazing allotment.

Jim Bark was a well-educated man for those days and came from New York to Arizona to seek his fortune. He was only 31 years old when he bought the ranch along with his financial partner, Frank Criswell. In 1904, Jim was elected to represent Maricopa County in the 23rd Arizona Territorial Legislature. About that time, he also helped organize the Arizona Cattle Growers Association (ACGA) and served as that organization's second president (1906–1908). He was a progressive rancher and immediately built a stone barn and hired a miner from the Silver King Mine for an entire year to blast out and improve springs and water seeps on the ranch. The barn and some of these drinkers are still in use today (see later figures). However, in the fall of 1891, an event occurred that was to reshape the rest of Bark's life.

Bark and Criswell registered these three brands and earmarks on April 30, 1897, as shown in the Livestock Sanitary Board, Territorial Brand Book of Arizona, page 77. The brands are (from top to bottom) JL, which was John LaBarge's brand; ML Connected, which was George Marlow's brand; and the Quarter Circle U brand. Courtesy of Arizona State Archives, Phoenix.

In October 1891, the "Dutchman" died in Phoenix. Bark had never met Jacob Waltz, who was referred to as the Dutchman, but after Waltz's death, Bark's ranch was inundated with gold seekers looking for the Dutchman's lost mine. The old wagon trail from Mesa ended at the Quarter Circle U ranch house, and this was the quickest way into the heart of the Superstition Mountains. Many of the visitors had maps and information, which they seemed willing to share with Bark—under the cloak of secrecy—because they figured he knew the physical terrain of the mountain better than anyone else. Bark kept detailed notes about all of the various and oftentimes strange visitors who came to the ranch asking for directions, horses, supplies, and sometimes guidance.

Bark eventually became so interested in finding the mine himself that he hunted for it, on and off, for more than 30 years. Bark sold the ranch to William Augustus "Tex" Barkley and his wife Gertrude in 1911 but continued to search for the mine with his partner, Sims Ely. He finally retired in 1928 to Pasadena, California, and died at age 78 in 1938. After he retired, he had his notes and collected stories typed, double-spaced, on letter-size paper. These are called the *Bark Notes* and comprise 174 pages. I, thanks to my son, Tony, have a copy of these notes. These notes provided the material for Sims Ely to author the book, *The Lost Dutchman Mine*, as well as a source of information for various other books on the Superstition Mountains.

Barkley Era

Tex and Gertrude Barkley, and later their son, Bill, and his wife, Betty, owned and operated the ranch from 1911 until 1967. In the 1920s, Barkley reported that they ran 5,000 head of cattle on the ranch. They were hard-working people who earned a living off this very rugged and harsh terrain. The harsh conditions are reflected in the faces of Tex and his working crew when they were photographed in the 1930s. See photos on the next page. Tex was honored in 1960 by being inducted as one of the original charter members into the National Cowboy Hall of Fame in Oklahoma City.

Tex Barkley at the Quarter Circle U Ranch, January 1932. Photo courtesy of Chuck and Judy Backus. The original photo is from Nancy Barkley McCollough.

Tex Barkley, second from right, and his cowboys at the Quarter Circle U Ranch, probably about January 1932. Photo courtesy of Chuck and Judy Backus. The original photo is from Nancy Barkley McCollough.

Tex Barkley, right, and Sheriff Jeff Adams, left, at the Quarter Circle U Ranch in 1932. The view is to the north with the Dacite Cliffs in the background. Photo courtesy of Chuck and Judy Backus. The original photo is from Nancy Barkley McCollough.

During the Barkley 56-year period of ranch ownership, there were major changes in Arizona that affected the ranching community. These changes included the creation of statehood for Arizona, the creation of the U.S. Forest Service, the creation of the Arizona State Land Department, a Wilderness Area designation for the Superstition Forest Service area, plus the human population explosion in the new State of Arizona. All these changes put major restrictions on the ranch operations in Arizona. In addition, the Federal Taylor Grazing Act of 1934 required that all ranches had to be separately fenced from each other if they had separate owners. Plus, the various federal and state government agencies established cattle-carrying capacity limitations and conditions for the users of the rangelands. As a result of these and other use restrictions, the conditions of rangelands in the United States have been continuously improving for the last one hundred years.

Early History of the Quarter Circle U Ranch

In Barkley's time, the Quarter Circle U Ranch included most of the county north of US 60, including Gold Canyon, and continued north on both sides of the Apache Trail up to what is now Canyon Lake. The eastern boundary went about straight north from Queen Valley, just east of Whitlow Canyon.

The ranch included all of the speculated sites for the Lost Dutchman Mine, and thus Tex was exceedingly popular with the various gold hunters, hikers, campers, and mining prospectors. He frequently packed people into the mountains and carried supplies to them. He was also involved in search and rescue operations, especially during the turbulent years of the battles between the mining camps and the frequent killings of people in the mountains.

A view of the historic Coffee Flat Corral, a.k.a. the Whitlow Corral, which was probably built by the Barkley Cattle Company or one of the earlier ranch owners. In the early years, these post and rail corrals were built using the local trees. Later, cowboys added wire fencing to replace the deteriorating logs. Photo by Jack Carlson, June 2010.

The most published episode of Tex's career was the infamous search for Dr. Adolph Ruth. Dr. Ruth turned up at the ranch in May 1931 with a map to the mine—one of the many, many different maps that existed. Tex was not anxious to pack this man, who was in his sixties and had a bad leg, into the mountains and said that he would wait until he had finished working his cattle. While Tex was away on a three-day trip, Dr. Ruth talked a couple of prospectors into packing him into the mountains. On Tex's return, he immediately went to the location of Ruth's camp at Willow Spring in West Boulder Canyon but could not find Ruth. This disappearance led to an unsuccessful search by the authorities. A few months later, a skull was found with two bullet holes through it and was identified as that of Dr. Ruth. Sometime later, Barkley found the rest of the body about a half-mile from the skull, with signs of bullet holes in it. The murderer, or murderers, of Ruth were never found. This search for Dr. Ruth made headlines in papers all over the United States.

Private Land Added to the Barkley Ranch

The Barkleys (Tex and son, Bill) owned the ranch from 1911 until Bill died in 1967. The Barkley Ranch ownership was established before the U.S. National Forest Service was created, before the National Wilderness Act was passed, and before any of the land surrounding the headquarters had become Arizona State Trust lands.

There was no ownership of private land associated with the ranch until 1947, when Tex applied to the State Land Department to purchase 140 acres around the ranch headquarters. However, before the time that Tex bought the land from the state, the state had already given a commercial lease of twenty acres to a movie company, where some of the early Western movies, including silent movies, were filmed. That twenty-acre movie set was located on flat land on the road to, and near, the ranch headquarters. Thus, when Tex bought the 140 acres from the state, he included the acreage around the headquarters and all the land up to and around three sides of the twenty-acre movie set, but not the movie set property.

Preparation for the Sale and Division of the Barkley Ranch

In 1955, Tex passed away, and his son, Bill, ran the ranch for 12 years until his own death in 1967. Then in 1967, Gertrude Barkley sold the two ranches and moved into a new home in Gold Canyon, where she lived well into her 90s. Mrs. Barkley, who was also a rugged cowhand, eventually became blind and was assisted by Nancy (Bill's daughter), and her husband, Ken McCullough of Gilbert.

For the sale, the U.S. Forest Service (USFS) required that the ranch, officially, be divided into two ranches—even if they both had the same owner. One ranch would be made up of U.S. Forest Service land, and the other would be composed of Arizona State Trust land, as well as some Bureau of Land Management (BLM) land. The western part of the Barkley ranch became the Superstition allotment of the U.S. Forest Service, and the eastern part, including the ranch headquarters, remained the Quarter Circle Ranch with a grazing lease from the Arizona State Land Department.

This division of the ranch was complicated. To hold a USFS grazing permit, the USFS ranch was required to own 10 acres of private land with facilities to manage cattle, which was usually adjacent to the USFS land.

A condition of the divided ranch sale stipulated that the buyer of the USFS Superstition allotment grazing permit would get 10 acres of the private land. And a legal agreement was made that required the buyer of the Arizona State Trust land ranch to drill a well and build cattle handling facilities on those 10 acres—all for the buyer of the USFS Superstition allotment.

Barkley Ranch Is Divided. John Cox and Joe Lamb

John Cox, who lived adjacent to the Forest Service land north of Apache Junction, agreed to buy the USFS Superstition grazing allotment from Gertrude Barkley, and that sale was completed.

However, when Joe Lamb tried to buy the Arizona State Trust land ranch from Gertrude Barkley, she refused to sell to him because he was well known for being dishonest. So, Joe went to a friend in Gilbert, Arizona, Alfred H. Nichols, to front for him—to buy the ranch. Nichols *officially* bought the ranch from Gertrude, with Gertrude holding a note for some of the money due.

Later, Joe Lamb bought the Arizona State Trust land ranch from Nichols and assumed the note to Gertrude and the obligation to drill a well and build cattle handling facilities on 10 acres of private land reserved for John Cox. For Nichols's services of fronting for him, Joe gave Nichols ownership of 30 acres south of the headquarters.

Joe Lamb proposed that the 10 acres for John Cox be located at the western end of the private land along the Quarter Circle U ranch road, just east of the junction with Peralta Road. He drilled a well there, but it was dry.

As a result of that dry well, Joe Lamb and John Cox modified their legal agreement so that Joe would drill a new well north of the Quarter Circle U ranch headquarters on 10 acres of Arizona State Trust land, which was not private land. The 10 acres were at the end of the road going north from the Quarter Circle U ranch headquarters. It was adjacent to the Forest Service land, and the grazing allotment could be serviced by that headquarters.

Joe drilled the new well, and it had water production that was acceptable to John Cox. Joe built corrals near the well and fenced 10 acres of Arizona State Trust land around these improvements. Joe then applied to the Arizona State Land Department to buy those 10 acres. However, the Arizona State Land Department had changed their rules and denied Joe's request to buy that land. That well and facilities, which we now call the Upper Corral, were still intact at the time that I acquired the Quarter Circle U Ranch.

Finally, Joe Lamb, John Cox, and the USFS agreed to use the 10 acres around the QCU ranch house and those cattle handling facilities as the base 10 acres of land to hold the Forest Service Permit. However, the agreement was that those 10 acres were to be officially owned (in title) by John Cox, the Forest Service holder, and Cox was to use the new facilities adjacent to the Forest Service land as his operational headquarters.

The legal agreement stipulated that Joe Lamb would share complete use of the current QCU headquarters with John Cox. The agreement was tenuous at best and could obviously lead to future disputes. John Cox was an honest man and continued to operate that way for many years—including after we had bought the QCU Ranch.

This awkward situation of having two ranchers using the same headquarters was not settled until the Forest Service canceled the grazing permit for the Superstition allotment. Additionally, when I was purchasing the ranch, I had to negotiate with the finance lender (that had financed the Forest Service rancher) for the purchase of the 10 acres that contained the ranch headquarters.

Joe Lamb and Guy Hill

When Joe Lamb acquired the ranch and the surrounding private land, he was interested in making a quick buck and moving on to take advantage of other people. Joe knew another rancher, Guy Hill, who was in trouble with the law in southern Arizona—accused of cattle rustling. Guy needed to get away, so Joe "sold" the QCU Ranch to Guy Hill, with Hill assuming the note due to Mrs. Barkley. The sale included the grazing permit on the Arizona State Trust land, the

10 acres of private land that the state required to hold a trust-land permit, and the right to use the ranch house and surrounding corrals.

Hill's purchase included 10 acres of private land across the creek from the headquarters ranch house. The ownership of the other 120 acres of private land was retained by Joe Lamb. He transferred 30 acres of that land to Nichols for fronting for him in buying from Gertrude Barkley, so he now owned 90 acres.

Joe later bought a ranch in southern Arizona from the widow of a rancher, with the widow taking ownership of these 90 acres of private land as a down payment on the ranch.

Meanwhile, Guy Hill went to the PCA (later known as Farm Credit) and convinced them to loan him the money to buy the Arizona State Trust land ranch (a.k.a. Quarter Circle U Ranch) from Joe Lamb. I do not know how much Guy paid, if anything, in down payment or in subsequent payments to PCA. He was running a livestock company that supplied animals (horses, bulls, cows, and calves) for various county rodeos and fairs. He was thus frequently away from the ranch for extended periods of time. He was also not making payments to the PCA.

Chapter 4

Backus Purchase and Early Operation

Purchase of Guy Hill's Foreclosed Ranch

A few years after I had talked with Dwayne Webb, who was a loan officer at the Production Credit Association (PCA), he called me and said that he had a situation that might be of interest to me. I thus again went in to talk with Dwayne—this was early in 1977. He said that PCA had a ranch that they had financed that met my requirements, but they were going to foreclose on the ranch. The name of the ranch was the Quarter Circle U Ranch.

He and I scheduled a time to ride the ranch on horseback, with me borrowing a horse from Dean Thompson. We rode all the way down Tule Canyon and back up No Name Canyon. Dwayne told me that if I would take over the debt and monthly payments due on the overdue note, they would agree to sell it to me. He said that I would, of course, have to buy cattle to turn out and that they would also loan me money to do that. The person who had owned the ranch was Guy Hill. I later found out that he was considered a dishonest person and a cattle thief.

I made a deal with Dwayne and Guy that I would agree to help them gather the cows from the ranch and would buy fifty cow/calf pairs from Guy. This arrangement would allow me to have some cows that knew the country. It would also provide calves that, the next year, could be heifers to add to the herd plus have steers to sell for income.

Judy and I made our first purchase of the ranch property from the PCA and Guy Hill in 1977. We got the 10 acres across the creek from the headquarters that was owned by Hill, the ownership of the Arizona State grazing lease, 50 cow/calf pairs, the Soap Pot brand, the remaining note to Mrs. Barkley, and all the agreements related to the sale of the Barkley Ranch.

The Quarter Circle U brand was registered by Joe Lamb, and that brand was transferred with the sale of the ranch to John Cox. The cattle that were on the ranch with the Quarter Circle U brand were sold before we bought the ranch.

We acquired the Soap Pot brand with the initial purchase of the ranch. The Soap Pot was a brand that Guy Hill owned but had not been using. We retained the Quarter Circle U Ranch name for the ranch on the private property. We soon paid off the note to Mrs. Barkley.

The USFS Superstition Allotment

In the mid-1980s, John Cox was retiring from ranching and talked with me about buying his USFS Superstition allotment adjacent to us that had been part of the QCU Ranch before it was separated by the USFS. After talking to John and the USFS office, I decided not to buy it. It would take too much of my time to manage the allotment for 200 cows in about 200 sections of Wilderness land. And I couldn't afford to hire and provide a house for another cowboy.

John then contacted Joe Lamb about purchasing the allotment. Joe talked the Farmer's Home Administration (FHA) into loaning him the money to buy John's allotment. Joe was also allowed to transfer more than $200,000 of other debts from ranching to add to the FHA loan.

The FHA is widely known as the lender of last resort. It was created during the Great Depression in the 1930s to keep farm owners from losing their farms with loans at especially low rates and long pay-out times. I'm not sure how Joe Lamb convinced the FHA into giving him this large loan with only the USFS allotment as collateral, but they did.

For several years, Joe's son, Kevin Lamb, operated the Superstition USFS allotment. I interacted with Kevin several times as he traveled through our headquarters to get to his allotment since the 10 acres at the Upper Corral were his allotment's operational headquarters and since Joe Lamb owned the 10 acres at our headquarters. My impression was that Kevin was as crooked as his father. It reminded me of an old saying: The fruit of a tree doesn't fall far from the tree.

The Lambs were behind on their payments to the Forest Service, so Joe talked to his friend Guy Hill about buying the USFS allotment and assuming his loan at FHA. This concerned me since I had been hoping to resolve the dual use of 10 acres at the headquarters by both Joe and me. Joe needed that 10-acre headquarters of private land to hold the USFS allotment lease. I asked my attorney, Richard Morrison, for advice, and he said that we could file a *Lis Pendens*, which is like a stay. That would put the sale of the ranch on hold until the conditions on the 10 acres of the previous sale had been satisfied.

The Trial in Florence, Arizona

Lamb, now on the other side of the deal, didn't want to go through the expense of acquiring 10 acres and have to get USFS approval for a different 10 acres of private land, so they filed a lawsuit against us to force the removal of this stay. We ended up going to court in

Florence. Since Richard was not a trial lawyer, another partner in the firm, John Gemmill, represented us.

John called Bob McConnell, who was the Assistant Attorney General of the United States in Washington, DC, to ask about the details of the purchase agreement. Since Bob still had family in Arizona, he offered to come to help with the trial. The trial lasted for a couple of weeks, with the judge finally calling it something like a draw—no decision, but meaning no sale—which was good for us. Of course, Judy and I had to pay about $50,000 in legal fees.

As an interesting note, when we were on recess during the trail, the secretary at the Pinal County Courthouse came in and said, "The White House in Washington, DC, was on the phone, asking for a Bob McConnell." Her face was as white as a sheet. It was Bob's boss asking him to return to DC. Another interesting note was that at another recess of the trial, I passed Joe Lamb in the hallway, and he stopped me to say, "In the old days of Patsy Cline, we would have taken someone like you out into the woods and just shot you."

Purchase of Headquarters 10 Acres

The FHA foreclosed on Lamb's loan, and sometime later, I went to the FHA Mesa office and asked about buying the USFS allotment. They said that I would need to assume Lamb's loan of about $500,000. I said that an allotment for 200 head with no private land was not worth that much, but they refused to consider less and did not consider any of my other suggestions and arrangements that I offered. I left their office knowing that I was probably the only logical buyer.

I was so disgusted with the FHA that I suggested to the USFS that this would be an excellent time for the USFS to just cancel the Superstition grazing allotment. They accepted my suggestion and canceled it. Thus, the FHA got nothing for their delinquent Joe Lamb loan.

After the foreclosure, the 10 acres of the ranch headquarters were now owned by the FHA. I inquired about buying that private land and was told that it was now being handled by the U.S. Treasury Department, and I would have to contact them at their Phoenix office. After several visits, we concluded that a price of $30,000 would be an acceptable purchase price, so I paid that amount for the 10 acres. By that time, I felt that I had paid for and bought that same 10 acres two or three different times, but I now owned the 10 acres that held the headquarters.

Purchase of Remaining Private Ranch Land

The 90 acres of private land (held by the southern Arizona lady who Joe Lamb had traded toward buying her ranch) had indeed been sold by her. The buyer was a partnership of 12 different people and couples. She had agreed to personally finance most of the price.

After I had met some of the partners, I was concerned about the quality of the note that the lady held because the partners who bought the property seemed to me to be untrustworthy. A few years later, that lady called me and offered to sell to me, for cash, the note for about 50% of its face value. After just buying the 10 acres of the ranch from PCA and Guy Hill, I did not have that much cash, so I asked my friend Professor Paul Russell to partner with me. He contributed cash, and I borrowed money from the PCA for my half of the purchase.

A few years later, as the partnership of the private landowners began to fall apart, they decided to divide the land into individual acres of ownership. Many of the owners ended up with 5- or 10-acre parcels.

Paul and I met with all of the owners. Several of them wanted to sell to us, but two did not. They wanted to build a store on their 10 acres so they could sell merchandise to all the people coming out to hike from the Peralta trailhead. They built a store with a bar and a large walk-in cooler. The business was not successful, so one of the owners just lived in the store.

A view to the north of the Lost Dutchman Mine store with the Three Sisters Rock formation on the horizon. Two men, Ernie Provence and Tracy Hawkins, who owned 10 acres of the original private land, built this store in the 1980s. The weather and insects took their toll on the structure, and its parts could not be salvaged, so it had to be torn down in 2018. Photo by Jack Carlson, 2006.

In 2011, we bought the store owners out, and Paul and I divided up the lands such that Paul then owned a complete 20 acres across from the ranch house and Judy and I owned 115 acres. The Superstition Area Land Trust (SALT) owned the remaining 5 acres of the original 140 acres of private land.

The whole time it took to acquire ownership of the 115 acres of private land (with SALT owning 5 and Russell owning 20) took more than 40 years. After Mike and Amy bought the ranch from us they were able to acquire the 20 acres from the Russell descendants and finally own all 135 acres. The ownership of the final 5 acres is presently held by SALT.

With all the agreements and confusion in the history of the ranch, we required all those previous agreements to be physically attached to our Bill of Sale and also to have all of the previous parties agree to sign a separate sheet that these agreements were still valid.

My Phoenix lawyer, who I knew from his student days, Bob McConnell (President of the ASU Student Association), collect and compile all the earlier agreements, with all parties again signing and verifying that the earlier agreements were still valid, and had those documents included in the new sales agreement. The new sales agreement turned out to be about one inch thick. Completing it demanded a lot of running around by Bob McConnell. The agreement thus turned out to cost a sizeable amount to Judy and me, but we felt it was required. Of course, the folks at Farm Credit were incredibly happy to have that done.

Early Evaluation and the NRCS

With all the caution and effort of preparing the sale agreement, we finally completed the ranch purchase. The sale included getting a few pairs of cows and calves, but additional cattle needed to be accumulated. That process began with the help of the PCA and continued for the next few years.

However, while buying cattle to start the process, I was interested in trying to improve the physical condition of the land. I made an appointment to talk to the Arizona State Land Department person in charge of Arizona State Trust land grazing leases. He said that the state had very few employees, and they just take care of the paperwork associated with the state trust leases. If I wanted advice and assistance on the improvement of the land, I should go to the Soil Conservation Service (SCS). It later changed its name to the National Resource Conservation Service (NRCS). This was perhaps the best advice that I was ever given.

The NRCS even had funds to improve the range conditions for which I could apply. NRCS expressed interest and suggested that they first do a detailed mapping of the ranch to assess the soil and range condition of the different areas and afterwards decide on what improvements were justified. That sounded very reasonable to me, so we started the process. I applied to NRCS for funding for that service, which was granted, and the process started.

First, Dan Robinet, from NRCS, and I had to ride the complete ranch in order to decide where the best places were to do sample collections for transect monitoring. Then we needed to sample the selected sites. Afterwards, we could talk about the improvements needed, such as where fences should be built and where water sites could be located to assure more complete utilization of the grasses.

A geologist from NRCS was assigned to do a geological map of the ranch, and the NRCS Range Con and I started the range-condition survey. This process took a few years to complete, but it helped me understand more about the plants that existed on the property as well as the grass-production conditions and potential of the various pastures on the ranch.

The way the NRCS works is the applicant writes proposals for materials for improvements—with the justification for the need; the NRCS staff evaluates the proposal and decides what, if any, of the materials they would supply. With that approval, the applicant buys the materials at their own expense and installs or arranges for the installation of the improvements. After the installation is completed, the NRCS staff has to inspect the installation before they reimburse the rancher for the materials. The NRCS does not pay for labor costs because those could be abused too easily. Once the project is completed and inspected by NRCS, the rancher gets paid. Because I had many, many projects and that they considered me trustworthy, they often would skip one inspection and inspect it on the next visit.

After the original ride of the entire ranch with Dan Robinet, it was clear the ranch needed fences down the main mountain ridges to separate pastures, and so they authorized the purchase of many miles of fence to get us started. The installation involved: first buying the posts and wire and getting them to the ranch; loading all of these long T-posts and heavy spools of barbed wire on several pack horses; getting the pack horses up through very rough canyons and to the ridges of the mountain ranges; driving T-posts into very rocky ridges; stretching the four strands of barbed wire between the posts; and finally inserting the metal stays through the barbed wire.

During the years that the NRCS survey was being conducted, Marvin Jones, the boy who lived at the ranch, and I started building fences. In order to properly manage the cows to improve the ranch, we needed to decide on a management plan to be able to move the cows in such a way as to graze different areas at different times. Since this required fenced pastures, pasture boundaries needed to be established. In the canyon areas, we could perhaps rely on cows staying in certain canyons. However, we wanted the cows to graze high on the sides of the canyons, but we did not want them to go across the ridge into another canyon. We thus needed to build fences down the ridges of the mountains. Since NRCS recognized that these fences were needed, they early on provided materials for Marvin and me to start the fence work. Even the ranch-boundary fences had not been maintained and thus had to be essentially rebuilt.

Transect Monitoring

One way of measuring the success of our multiple pasture creation was to establish many monitoring sites. The NRCS personnel and I selected the locations and established at least one monitoring site per pasture. The sites were chosen where there was use by cows, perhaps halfway up the side of a mountain with multiple plants and bushes, but

View to the east with John Patton, right, holding the transect device and identifying the plants. He is calling out the plant names to Chuck Backus, left, who is recording them on a tally sheet. Note that the vertical handle for the transect apparatus broke, so we found a stick to replace it. Photo by Jack Carlson, January 2006.

not in a lush area. But not in areas heavily used by cows, such as near a water hole. The sites were to have vegetation that was representative of the vegetation in that pasture. The plan was to monitor and take measurements at the study sites annually or biennially. The starting point of each transect site was established by erecting a stone cairn or monument several feet high so that it could be seen from a distance.

A close-up view of some of the plants observed in the transect study. From the left: hopseed bush, slim tridens perennial grass, and hairy grama perennial grass. Photos by Jack Carlson, January 2006.

This study was called *transect monitoring* because it followed an imaginary line at the same elevation around the slope. At the starting stone cairn, a standard square-shaped metal form, as prescribed by the monitoring technique, was placed on the ground. Then all the different species of living plants in that square were recorded on a tally sheet—grasses, shrubs, mosses, etc. Only the plant's name, not the number of plants, was recorded on the tally sheet. We then took two more steps in the same direction, placed the metal square on the ground, and recorded all the species therein.

We continued the transect around the hill for 50 sample recordings. At that point, we took two very large steps down the hill from the first transect line, did the same monitoring sequence of 50 samples going back along the slope, and ended up about eight feet below the starting stone cairn. This obviously supplied 100 samples of the data. From that data, we counted the times that a certain species occurs—say, 25% of the samples contained plants of sideoats grama. If we returned to this site, repeated the same procedure every other year, and complied data for several years, we could easily determine which plants were increasing or decreasing at that site.

For our data to be meaningful, we needed to accumulate multiple pasture readings over many years to see the trends in a pasture. Since we have been taking these measurements for more than 40 years, we can see the health of the plants improving. When we were setting up the original transects, we tried to choose a site near the top of the pass on the trail going from Tule Canyon into No-Name Canyon. We set up a site just on the No-Name side of the pass. After taking one pass of 50 recordings, we observed only one perennial grass in one sample. Thus, we decided that it was not a good location for monitoring. About 15 years later, the NRCS staff suggested that we go back and read that same site again—just for the heck of it. We did, and we observed perennial grass in about 100 of the samples. We were very happy, especially me, but it was not an official observation since the site had been discarded after the original reading. It showed that our efforts to improve the pastures were producing the desired results.

The readings of these sites over the years vary dramatically and show the impact of rainy periods and drought periods—more plants in the rainy years and fewer in the drought years. The data shows an overall increase in the various grasses and desirable plants, which indicated that our management plans were very effective. However, they also showed the dramatic impact of droughts.

It also showed that it is desirable and beneficial to have a lot of browse, such as bushes and trees, on the ranch for cattle to graze. The bushes are, of course, affected by drought but still remain sort of green and edible by cattle. Their hardiness is due to the deeper roots that bushes have, so their sensitivity to short periods without rain is low. The edible trees are extremely useful for cattle in this type of country, but it also depends on the type of tree. One time, we took samples of cow pies over a large area, and the NRCS itemized the percentage of plants in the samples. The cow pies contained a large percentage of bushes that the cows had been eating, and thus, it confirmed that bushes are an important food source.

The monitoring transects have been very useful as a way of obtaining and recording range data for use by the ranch, the ASLD, and the NRCS. Educationally, transect monitoring has been used as a training method for new employees at the NRCS, Arizona State Range Conservationists, and ranch personnel. Oftentimes, visitors would come along and be surprised at how technical and thorough cattle management and ranching have become. Of course, after a few years of experience, one can just ride through a pasture on horseback and assess the condition of a pasture. It is very satisfying to me personally to see that, after managing this ranch for 40+ years, the range has improved for the cattle, the ranch, and the rangeland.

Improvements at the Headquarters

When the ranch was purchased in 1977, the improvements at the headquarters included the ranch house originally built by Matt Cavaness in 1876, the stone barn built by Jim Bark in 1891, and a small wooden corral that surrounded the barn, which was connected to a large wire fence. There was also a double-wide house trailer brought in by Guy Hill about 1973. However, I later found out that the house trailer was on land that was owned by Nichols. When I learned that Nichols had passed and his son had inherited the land, I located the son and negotiated to buy all of the 30 acres he had inherited that were located on the ranch. This was especially beneficial for the ranch operation since it also included the Lower Well.

A view to the southeast of the stone barn and outbuildings in 1962. At that time, there were only a few wooden corral fences. Courtesy of Keith Ferland and the Kollenborn Collection.

Tom Kollenborn in 1977 is standing in the area where we built the corrals and pens. To the left and behind Tom is a standalone squeeze chute, not connected to any of the corrals. To the right of Tom, it looks like there is a loading chute connected to the wooden corral by the windmill. Courtesy of Keith Ferland and the Kollenborn Collection.

Sketch and watercolor by Tom Kollenborn about 1959 showing the bunkhouse and the surrounding structures. The small building on the left is the original tack room. Courtesy of Keith Ferland and the Kollenborn Collection.

In 1977, the ranch manager lived in the ranch house, and we occupied the double-wide trailer. Plans slowly started on building a set of corrals. They were built out of 1-inch-thick steel sucker rods. Corral designs were studied, such as those by Temple Grandin, and materials were purchased gradually as funds became available. See pictures in my booklet titled *Cattle Operations at the Quarter Circle U Ranch* for the story I wrote in 2010 for a presentation on ranch-operating facilities. The steel corrals were originally painted white. However, because they stood out so vividly when viewed from the trails in the Wilderness Area, we later painted them a dark red-rust color to blend into the terrain.

Fire at the Stone Barn

I know that we started to build or rebuild the corrals when Henry and the boys were still living at the headquarters. Merlin Jaeger worked as a welder at the Air Research Corporation in Phoenix, and he taught Marvin how to weld. I remember one time when Henry's boys, Marvin and Brian, were welding near the stone barn on a Saturday morning. I came out late that day, and I brought lunch. They stopped welding and came up to the ranch house to eat lunch.

About the time that we finished lunch, we heard a very large explosion. We rushed outside to see the wood building adjacent to the stone barn completely engulfed in flames. We could not go very near the barn, but we were able to rush down and remove the welder and the gas tanks from the burning building. Since the wooden building was close to the stone barn, the roof of the stone barn and a wooden shed on the other side of the stone barn all burned. We were able to hose down the areas around the barn so that the manure in the side corral and in the main corral, in front of the barn, could not burn.

There was an assessment after the fire finished burning. The boys had been welding on the steel sucker-rod fence from the wooden building as I drove up. Evidently, the molten metal from the welding had dropped on the ground beneath and started a fire in the manure just as they walked away. It was good that neither the wooden building nor the stone barn had any valuable goods in them. The stone walls had been blackened somewhat from the fire, but they were structurally fine. The explosion that we heard was from a tire on the welder that had exploded. The boys felt so extremely bad about causing the fire that I did not have to scold them.

After the fire debris was cleaned up, Merlin Jaeger and I sat down and discussed how we could build it back in a way that would be more useful to the ranch. Merlin was a square dance friend that helped me every weekend at the ranch until after we both retired and then every day after that. What we decided to do was to build a large concrete slab on the south, well side of the stone walls for a hay-storage location for feeding animals in the corrals near the barn. We would then build a roof truss over the entire barn and concrete slab.

We built the forms for the entire hay-storage floor area and then scheduled a concrete truck for a Saturday with a load for the estimated area and depth of the concrete. As I recall, we also asked Bill Smith's son to join us for pouring and finishing. The concrete pour went well, so Merlin and I designed and started to build the 2 x 6 trusses over the stone barn and the new concrete slab. The south wall of the stone barn served as the center of the roof and supported the trusses of the roof. We installed steel posts from the ASU farm on the south side to hold the roof tin. The north side of the roof was supported by the north side of the stone wall of the barn. After Merlin and I built the trusses, we added the 2 x 4 supports for the tin sheets of the roof. It turned out very well for being a walled inside storage area plus an outside hay barn. We then built a concrete bunker feed trough on the south side of the new barn and the corral by the water well.

After the roof on the stone barn burned in the welding accident, Chuck and Merlin designed a new structure that added a hay shed to the west side of the stone barn. The new roof covered both the old stone barn and the new hay shed. This is a view of the south side of the structure looking to the northeast. Photo by Jack Carlson, September 2007.

Evolution of Branding Methods

When we bought the ranch, there was a wire fence west of the stone barn with a small holding area and a wooden loading chute on one side. Originally, all the branding we did started with all the cows and calves in the large fenced wire pen. The calves had to be roped and dragged to an open fire, which held the branding irons. Of course, the mothers didn't like that and often followed the bawling calves to the fire and interrupted the branding process.

After dragging the calf to the fire, it had to be hand-flipped to its side and held or have its feet tied. Then the calf had to be held still while another cowboy took the iron out of the fire and applied it to the proper side and location on the calf. Of course, the mother cow would often try to keep the cowboys from hurting her calf by being obnoxious and interfering with the working cowboys.

Before we built the corrals and pens for sorting, the cowboys had to rope the animals and drag them to the branding area. In a March 1959 photo, ranch hand Tom Kollenborn is shown roping a cow in the small corral by the stone barn. Courtesy of Keith Ferland and the Kollenborn Collection.

Typical working cowboys love to rope calves in a herd of cows and calves and then drag them to the fire—it's just a fact of life. For myself, I was looking forward to a time when we had multiple steel pens that allowed us to separate the calves from their mothers. Then, without their mothers, we could put the calves in a pen and catch, throw, and hold them down while they were being branded. That became possible when we installed multiple steel corrals.

In a typical wood fire for branding, it is always a problem to keep the fire going and keep the iron in a part of the fire to do the proper and uniform heating. Thus, the brands did not always look uniform on the side of the calf because the iron was not evenly hot or not properly applied.

View of the solar panel array and the original small barn that housed the solar electric inverter and storage batteries. Photo by Jack Carlson, October 2007.

I designed and made a propane-fired branding apparatus that had a separate cast-iron holder with a propane flame so that a constant temperature could be maintained on the iron. We went with that for quite a while until we had solar-generated electricity wired to the working area and then changed to an electric branding iron with a constant temperature. I often told people that my cattle were all solar-branded.

Building the Tack Room

A much larger tack room was needed, and I decided that, for convenience, it should be built just north of the ranch house.

Someone had given Lee Thompson a set of about 12 exceptionally large roofing trusses that had an off-center peak to them. He just had them stacked on the ground in one of his pastures, exposed to the weather. He offered them to me for the new tack room. Thus, I designed a new tack room based on the size of those trusses.

The base of the trusses was 40 feet long, with the off-center peak at 10 feet and a steep side on what became the open-air front side of the tack room. I made the front wall of the tack room at the peak. The entire front area was left open as a staging space for saddling horses. A horse hitching rail was built across the entire length except at the center, which was at the door into the tack room. Also, we wanted to leave the north end of the building as an outdoor storage area. Thus, the inside area of the tack room was to be 44 feet by 18 feet, with a poured concrete floor.

Arkie Johnson was the ranch manager at the time, so we asked his brother, Jack, to help us pour and finish the floor. Jack Johnson was working for a concrete company. The distance across the tack room was 44 feet, so support posts would be needed every 12 feet to help support the roof trusses; thus, one would expect that those posts in the middle of the concrete floor would potentially cause cracks in

An aerial view of the Quarter Circle U Ranch shows the headquarters in the lower center. The ranch is mostly on Arizona State Trust land and borders the Superstition Wilderness (top half) for 10 miles. This photo was a Christmas gift from Amy and Mike.

the floor, and they did. The tack room was going to be too dark, so some of the corrugated roofing sheets were installed with transparent plastic panels. This facility was built by cowboy labor again and has served the ranch very well.

Since I was still working at ASU five days a week, I could only work weekends at the ranch. It was clear that all the improvements needed at the ranch were a multiple-year project. I hoped that I could complete the improvement projects before I retired from ASU and then devote all my time to cattle ranching.

View of the inside of the tack room with the poured concrete floor. Saddles are stored on the left side of the room, and the packsaddles are stored on a cowboy-engineered support pole in the center of the room. Photo by Jack Carlson, December 2007.

Chapter 5

The Beef Co-op and My Plan to Expand

The Arizona Natural Beef Co-op

In the mid-1980s, a group of us Arizona ranchers decided that we should create the Arizona Natural Beef Cooperative to sell a labeled product called Arizona Natural Beef to local Arizona markets. None of us were making much money and thought this co-op might help us earn more by eliminating all the middlemen.

We invited an Oregon rancher, Dr. Hatfield, to talk to us about creating a cooperative. He was a veterinarian who had organized and created a cooperative in his state. We followed his guidelines and formed the Arizona Natural Beef Cooperative. I was an officer.

We had members who had degrees from the Ag College at the University of Arizona (UA). The UA was the land-grant school in Arizona—thus, it had a farm, a feedlot, and a slaughter facility for educational purposes. One of our co-op members owned, with his father, a feedlot in Casa Grande, which we used, and we arranged with the UA to slaughter our cattle. We then contacted several markets and arranged for them to buy Arizona Natural Beef—beef raised in Arizona.

We operated this co-op for several years, into the mid-1990s. I was one of the last members in the co-op. We finally dissolved it about 1995. We did learn a lot about the US cattle industry: the operations of many of the ranches in Arizona, the operations of feedlots, the finishing of cattle for slaughter, the processing of beef, the transportation of beef products, and the efficiency and competitiveness of the entire US beef industry. However, we didn't make any money! But I felt that I had learned a lot from my entire co-op experience.

My Bad Decision

During the early 1990s, when we still had the co-op, the country was going through an economic depression, and I thought it might be a good time for me to expand my ranching operation. My university income and the mobile home park were not directly affected by the depression. However, in the early '90s, some of the banks were closing and/or consolidating, lending companies were in rough times, loans were hard to find, and interest rates were high.

I talked to several ranchers at the Wilcox ACGA summer conference. A friend said that the Goswick Ranch, adjacent to him near Meyer, Arizona, was for sale, and he gave me a phone number to call. I called and made an appointment with Rink Goswick for Judy and me to go up to inspect and tour the ranch.

Unfortunately, the outfit was larger than we expected, and we thought the price would be too high. They had a 350-cow USFS allotment and a lot of private land. The Goswicks agreed to carry over half the cost of the ranch if we paid about half of it down and cash for the cattle. We secured the funding and proceeded. We'd invested in improvements and operated this ranch for three years when the bank called our entire debt due because of the financial condition of the country. Thanks to Louis Maxy, our loan officer and friend, we arranged to borrow the money. We finally paid off that ranch debt

Quarter Circle U Ranch headquarters looking northeast toward Miner's Needle. Shown left to right: manager's house (removed in 2018), solar panels, solar electric power station, bunkhouse, tack room, branding station, hay barn, and the stone barn on the far right. Photo by Jack Carlson, April 2008.

when we got the final settlement for selling the trailer park a few years later. It was the worst decision of my life to buy this ranch!

At a much later date, in about 2002, I decided to buy another ranch. But that ranch was at a much higher elevation, about 7,000 feet, and would be compatible for use as a "summer pasture" in conjunction with the Superstition Ranch as the winter pasture. I could then double the number of mother cows in the herd.

Chapter 6

My Ranch Managers

Word gets out in the cattle and cowboy world about ranchers who may be looking to hire a cowboy. I had several people offer to live and work on the ranch. Many of the miners from Superior had horses and would help surrounding ranchers on weekends. Several helped Guy Hill and me when we were gathering Hill's cattle when I bought the ranch. Those riders included Chuck Sanders, who I hired as my first ranch manager.

My ranch managers in the early years were Chuck Sanders, a man nicknamed Rabbit who lasted only three months, Henry Jones, Arkie Johnson, and Herb Herbert. In more recent times, my managers have been Howard Horinek, Dean Harris, and Jordan Selchow.

Chuck Sanders

Chuck Sanders had previously worked at the mines but drove a road grader for Pinal County. He worked alongside Jimmy Gillette, the brother-in-law of my ranching neighbor, Billy Martin, and his wife, Teta. Jimmy was Teta Martin's brother.

Billy later told me that the surrounding ranchers were going to take up a collection to help me buy Guy Hill out if I needed financial help. No rancher wants a cattle thief as a neighbor. I also thought it was an advantage to have the county road grader work at the ranch, especially since his area of responsibility included Peralta Road. Chuck soon added the ranch road to his service area. I agreed to pay him $400 per month to live at the ranch headquarters.

Meanwhile, Chuck had interactions with a Montana rancher who offered him a job. Thus, after two years at the QCU, Chuck went to Montana. I hired a man nicknamed Rabbit from Superior at Chuck's suggestion. He had helped Chuck at the ranch. But that lasted only about three months before I fired him for not doing what I had asked him to do.

Henry Jones

I had ridden into Castro Cabin the week before I fired Rabbit and found Henry Jones and his two boys and girl cooking some food over an open fire behind the cabin. He didn't have a job and seemed to be in dire straits.

Right after I fired Rabbit, Henry stopped me along the road and said that he would like to live at the QCU ranch house and that his two boys, who were maybe 9 and 11 years old, would do ranch work for me, and he would take care of the headquarters at about half the price I was paying Rabbit—thus $300 per month for Henry. Henry was not very tall and weighed more than 300 pounds. I said we could try it for a while. Henry and the boys were there for the next five years—Marvin for six years. Marvin was the older boy, who was maybe 11 years old when I hired his father.

Marvin and I built many miles of barbed-wire fences, mostly down the ridges of mountain ranges, during the six years he was there. These were prescribed by the range plan that came out of the National Resources Conservation Service (NRCS) range study that I had initiated. The plan required many miles of fences to divide the ranch into pastures that allowed the proper rotation of grazing to improve the range conditions.

Hank Jones, working as a wrangler for the OK Corral Stables, is packing a group of campers from Peters Mesa to the Tortilla Trailhead in the Superstition Mountains. When he helped Chuck build fences at the Quarter Circle U Ranch, he used his middle name of Marvin. Photo by Jack Carlson, April 2005.

This requirement to divide the ranch into pastures was very apparent from the start, and so NRCS gave me money to buy lots of barbed wire, T-posts, and stays early on so we could get started before the study was completed. Marvin and I built most of the fences. This obviously required us to use horses to pack wire, T-posts, and stays to all of these ridges and other places where fences were required. As one would expect, these packing exercises led to many difficult and dangerous situations, but Marvin and I were successful! I have no idea how many miles of fence we installed, but it resulted in a range that has several pastures and many options for cattle management. The longest fence was on the ridge going completely between Tule Canyon and Fraser Canyon—four aerial miles long.

Henry Jones, Marvin's father, always stayed at the ranch house and was thus a good caretaker. He did have a taste for whiskey. He also spent most of the summer months without a shirt on. With his very fat and topless body, covered with old burn scars, he presented a rather grotesque figure. I always presumed that, being unemployed with several children, he was on some type of government welfare program. In the later years, he was bribed by people wanting access through our private property to drive up to the Forest Service Wilderness property line. I understand that a bottle of whiskey was usually the price for crossing through the ranch. In later years, this became more of a problem, despite Marvin being so helpful to me.

Arkie Johnson

When I first bought the ranch, I found out that a man named Arkie Johnson and his family had lived in the ranch house the year before we bought the ranch. I found a box with a canceled check with his name on it. The year before, Guy Hill had rented the ranch house and the corrals to Arkie for the purpose of running a riding stable to rent horses to people. Arkie owned the horse stable with a man named Bud Lane.

When we bought the ranch, Arkie and Bud bought a stable in Apache Junction (AJ) and ran their operation out of it. Bud Lane had worked for the Barkley Cattle Company and knew the ranch well. Arkie and Bud had one of the few, maybe three, commercial permits issued by the Forest Service for guiding horseback rides into the Wilderness Area. Their particular niche was to pack miners to their claims and also to pack supplies to the miners operating in the Superstition Wilderness. This permit was good only through 1983. The permits were allowed for only a fixed amount of time after the Forest Service designated this part of the Tonto Forest as a Wilderness Area. I got to know Arkie and Bud fairly well since I allowed them to pack their people and supplies through our ranch headquarters.

As Arkie's Forest Service permit was scheduled to end on December 31, 1983, Arkie started to talk to me about the prospect of returning to the QCU ranch as a ranch manager. I knew that Marvin was planning on graduating from high school in May 1985, so I decided that I needed to consider Arkie's offer. I think that Henry was also thinking about this and talked to someone in Tucson about a job. I think that he thought about leaving, or asked about leaving, after Marvin's junior year of high school. I talked to Arkie about what he had in mind about housing, and he said he would plan to move his double-wide trailer house out to the ranch. I asked Arkie, Marvin, and Henry about Marvin living in the ranch house and finishing out his senior year at Apache Junction High School. That is what we did, and Marvin thus became the first one in his family's history to ever graduate from high school! I was very proud of him. As I recall, he then attended a one-year school for diesel mechanics. Arkie moved his double-wide trailer south of the ranch house in 1984 and became the ranch manager.

The year after Arkie started at the ranch, the grazing people at the State Land Department called to ask if I would be willing to add two more sections of State Trust lands to my allotment. They said that they were canceling the permit on two sections of state land adjacent to the Superstition USFS lands north of Gold Canyon. As it happens, these two sections were held by Joe and Kevin Lamb and that the Lambs had turned out about 1500 steers on these two sections—yes, fifteen hundred. The worst part was that these two sections did not have any fencing along the two miles of their southern border. Those steers had run all over the golf courses and houses in the Gold Canyon area, thus warranting the cancellation of their grazing lease. I agreed to not put out any cattle until after we had installed those 2 miles of fence. With the fences in, I could then run an additional 20 head of cattle. Since I was reconsidering how to rotate the herd, I thought that I could use this pasture as a bull pasture to isolate our bulls from being adjacent to any pasture that contained cows.

During the time that Arkie was ranch manager, I leased out pasture to other cattle owners when we had a wet spring. One of those lease owners offered Arkie a job to work for him.

Herb Herbert

When Arkie left, I hired a man called Lee that Arkie knew. Lee had a complicated past, and soon afterward, he was shot and killed at the ranch. After Lee was killed, a person who had been riding with us, Meredith A. "Herb" Herbert, volunteered to live at the ranch house until we found another person. Herb ended up staying for 12 years as our ranch manager. Herb was always there and rode with us on weekends all those years. He was a great help for me to adjust to becoming a rancher. After Herb left, I started a search for a new ranch manager, and I was fortunate to have Howard Horinek apply.

Howard Horinek

Howard was the longest-serving and best ranch manager that I had. He had served in the US Army overseas and had been a ranch manager at several ranches in other U.S. states. He was a hard and consistent worker and knew what to do to operate a ranch. He was 11 years younger than me. Directly before coming to work for me, he owned his own house in the town of Superior and had been shoeing horses for ranchers in the surrounding region for the last several years. After I bought the Northern Ranch, he stayed at the home ranch and spent the time making several improvements to the Superstition Ranch. He and Jack Carlson installed many, many water systems all over the Superstition Ranch that made the ranch more usable and flexible. See the later chapters of this book where I discuss their work.

After many, many years of faithful work, I noticed that Howard was slowing down, and we had to hire extra hands to get the general work done. I still remember having a discussion with Howard about his possible retirement time. He said that he hadn't thought much about that. I said that I thought it was about time for him to start thinking about it and that I wanted to have him help me choose his replacement over the next year. Over the next few months, we interviewed several interested parties. It turned out that Dean Harris, our cattle truck driver, had sold his truck to spend more time at home with his wife.

Former ranch manager Howard Horinek at the Lower Well pointing to a log that was washed up on the bank when Navajo Wash had recently flooded. Cow dogs Spike, left, and Gus, right, are coming by to see what we are doing. Photo by Jack Carlson, August 2005.

Former ranch manager Howard Horinek branding a calf with the Soap Pot brand. He is looking at his watch on his left hand and timing how long to apply the branding iron—about 40 seconds. Photo by Jack Carlson, October 2007.

Former ranch manager Howard Horinek and Judy and Chuck Backus.

Dean Harris

Dean was born and raised in England and drove cattle trucks all over England and continental Europe. I cannot imagine driving those big trucks after seeing those narrow streets in England and Europe. In his twenties, he married an American woman named Kris and moved to the United States. He bought a US-style cattle truck and started hauling cattle in Arizona. I got to know him when he was hired to haul cattle between my two ranches. One night the two of us were sitting on the front porch of the bunkhouse talking, and he said that this was the only ranch in Arizona where he might consider living.

When Howard and I started a search for his replacement, I gave Dean a call. Judy and I met with Dean and Kris to discuss their interest. It turned out that both of them were interested in the job because it allowed them to live more remotely. Dean was obviously very familiar with cow behavior since he had hauled them for so many years.

Former ranch manager Dean Harris at the headquarters hay barn corrals. Photo by Jack Carlson, December 2017.

In a seemingly unrelated activity, our daughter had started a business of hosting weddings on some of the private open land at the ranch. It turned out that one of the wedding parties (from in town), where people had become very drunk, had decided to visit the wedding site—the ranch. Several cars came out to the ranch house late at night. With all the noise they made, Dean came out of his trailer with a pistol in his hand. This scared the partyers, and the next day they called Amy and canceled the wedding at the ranch. After hearing about this episode from Amy the next day, I went out to the ranch. Dean met me with the announcement that he and Kris were going to move back to Tucson.

Jordan Selchow

I got to know Tiffany Selchow because she worked for the Arizona Beef Council and would bring large groups of students or other groups to the ranch to show where meat originated. When Tiffany heard that Dean and Kris had announced they were leaving, she and her new husband came to our home in Gilbert and said they were interested in moving to the ranch. Her husband, Jordan, had graduated from the University of Arizona with a degree in Ag Education and taught at Gilbert High School for a couple of years but decided that he did not like his job. They both wanted to be the ranch managers. Tiffany would keep her job in town and commute from the ranch. We agreed to a salary, and they took over as ranch managers. That is the first time that we had managers who were college graduates. It worked out very well, and they are still the ranch managers—even after Mike and Amy bought the ranch.

Ranch manager Jordan Selchow is leading a packhorse to Coffee Flat Spring in the Fraser Pasture to make repairs on the spring box and to remove some airlocks in the piping going down to the drinkers and storage tank. Photo by Jack Carlson, July 2019.

Chapter 7

Ranch-Related Events

A&M Universities and the Creation of ASU

Land Grant Universities were established in the 1800s by the federal government during President Lincoln's term. These universities ensured that the residents of all the states and territories had an opportunity to study in the fields of agriculture and mechanics (engineering and technology). They were called A&M universities. In the Arizona Territory, the University of Arizona was established in Tucson to serve this A&M purpose.

Since Phoenix and central Arizona had abundant water and a canal system to distribute that water, the area developed as an agriculture center with a large population. The people living in the Phoenix area demanded that the Territorial Legislature establish a college to serve their community. Thus, in 1885 the Territorial Legislature provided funding for a teacher's college, which was named the Territorial Normal School at Tempe. However, as more industries started moving into the Phoenix area, those industries insisted that the legislature allow technology to be taught at the teacher's college.

The local industry and community continued to press for a broadly based university. Thus, Arizona State University (ASU) was authorized in 1959. Dr. Lee Thompson was hired from Texas A&M University to start the College of Engineering, Technology, and Architecture. Lee was raised on a ranch in northern Arizona and had gone to Texas to get his PhD in engineering. Afterward, he stayed at Texas A&M as a professor.

When I came to ASU in 1968, the College of Engineering and Applied Sciences included the School of Agriculture and the School of Technology. By then, architecture had formed its own college. A prominent farmer in Tempe had donated a large 320-acre farm to ASU, and thus the School of Agriculture had its own university farm. The school grew into prominence with the support of all the farmers in the Metropolitan Phoenix Valley.

Meanwhile, the direction of research I was pursuing at ASU was in photovoltaics, which is the direct conversion of sunlight into electricity. In 1982, I obtained a major contract of $200,000 from the National Science Foundation. This was a joint contract with a private company in California, and it was the largest grant that ASU had processed at the time. A few years after I received this funding, Dean Thompson announced that he was retiring as the Dean of Engineering and asked me to chair the committee to select his replacement. We hired Dr. Roland Haden. Soon thereafter, Dean Haden asked me to become his assistant dean for research.

Equipment Purchased from the Closure of the ASU Farm

The ASU farm had grown to the point that it had more than 100 horses for the popular riding courses, a dairy of about 150 cows, more than 200 chickens, and various other facilities to support farming courses taught by the Ag School that was in the College of Engineering and Applied Sciences. The related facilities included many holding and working pens, plus many riding lanes and pastures.

When the Board of Regents that oversaw the three universities in Arizona obtained a majority on the board from Tucson, the board told ASU to close the ASU farm and sell the 320-acre farm. The UA wanted to appear to be the primary Ag school in the state. Since I was the Assistant Dean for Research of the college that contained the Ag farm to be closed, I went to Dean Haden and suggested that rather than selling the land under the ASU farm, he should request that the regents turn that land into an ASU industrial research park. Besides providing continuous funding to ASU, the research park could provide an opportunity for ASU professors to interact with industrial researchers. Dean Haden pursued that suggestion and received approval for the research park, so we didn't have to sell the land.

The decision to close the ASU farm necessitated that all of the animals and facilities on the farm had to be sold. ASU thus hired a private company to organize an auction to sell everything on the land. That company hired three auctioneers who, on a well-advertised day, sold all the animals and equipment that people would bid on. There were no bids for many pens and buildings.

After the auction, Lee Thompson and I decided to talk to the people that ran the auction. We told them that we both owned ranches and had cowboys and friends who could remove all the items that were left. We asked the auctioneers to set a price for all of the remaining items. They set a price and said that everything had to be removed within the next three months because the bulldozers were scheduled to come in and level the land for the creation of the research park. We immediately paid the price and instructed our cowboys and other willing friends to help us remove the equipment and facilities. We subsequently hauled all that dismantled equipment to our respective ranches.

All of the above happened after I had owned the ranch for about five years, and I was ready and interested in making improvements at the ranch headquarters. I had just hired a new *real* cowboy/ranch manager named Howard Horinek. If one does not inherit a ranch, then it is difficult to accumulate enough resources to operate a normal-sized Arizona cattle ranch, so the closure of the farm was a true blessing for me because I could use all of these pipes, sucker rods, posts, roofing, and used building materials for new corrals and buildings at the headquarters.

More Construction Material Acquired

Sometime later, in the early 1990s, when I was working at ASU East, while driving down Williams Field Road, I saw a cow dairy being dismantled. I stopped and asked who was in charge, since there was a large hay-storage structure that remained after the tin roof had been

Eddie Christopher is delivering a squeeze of hay to the hay barn at the Quarter Circle U Ranch headquarters. Photo by Jack Carlson, December 2017.

removed. I arranged a deal that we would remove the structure and keep all the materials. It was built out of very tall telephone poles, and the roof had been supported by long 2 x 4 and even larger support beams. With precarious climbing and beam removal, my cowboys and I got the materials down and moved them to the ranch. Since the telephone poles were so tall, we were able to cut them off at ground level and carefully get them down. Somewhat later, we re-installed these poles at the ranch. I bought new roofing metal, and with the tin from the ASU farm, we covered the structure. We installed this hay barn on the west side of the big corral and later built a huge concrete manger (feeding trough) on the eastern edge of the corral for ease of feeding directly from the hay barn. The ranch's metal corrals and cattle-working areas were built over several years as I accumulated material from various sources. When Judy sold Lee Thompson's acreage near Lakeside, I bought some of his equipment. The ancient calf-rotating squeeze chute, which is still in use at the ranch, was from Lee's acreage.

Abundance of Wildlife on the Ranch

The ranch adjoins the USFS Superstition Wilderness Area along a 10-mile-long common border, and it also borders other land with Arizona State Trust land grazing leases. As a result, this undeveloped ranch land supports an abundance of wildlife.

A local man, Bobby Beeman, was interested in the wildlife that was present at the ranch and offered to set up motion-activated cameras to record various wild animals that came to water at the drinkers we had installed for cattle watering. The following pictures show the variety of wild animals that existed on the ranch—thanks to Bobby Beeman. The last two pictures are just some of the reptiles and snakes that we personally encounter at the ranch. The western diamondback rattlesnakes with a black and white tail are quite common.

A deer and her fawns at Turf Spring in the Southern Pasture. Photo by Bobby Beeman, August 2011.

Nighttime view of a herd of javelina (peccary) at the water tank in Tule Canyon. Photo by Bobby Beeman, May 2011.

Bighorn sheep at the water tank in Tule Canyon. Photo by Bobby Beeman, June 2011.

View to the northeast of Miner's Needle of deer at a Horse Pasture water trough. Photo by Bobby Beeman, September 2011.

View to the northeast of Miner's Needle of a deer and a bobcat at Bull Spring water trough in the Horse Pasture. Photo by Bobby Beeman, September 2011.

Black bear at the water tank in Tule Canyon. Photo by Bobby Beeman, May 2011.

View to the northeast of Miner's Needle of a black bear climbing out of the Bull Spring water trough in the Horse Pasture. Photo by Bobby Beeman, April 2010.

CR 7.7. Mountain lion at the water tank in Tule Canyon. Photo by Bobby Beeman, June 2011.

Gila monster on the trail in Bluff Canyon near the junction with Little Bluff Canyon. Photo by Jack Carlson, April 2005.

A western diamondback rattlesnake at Coffee Flat Spring. Photo by Jack Carlson, June 2010.

Events for Ranch Family and Guests

We have been fortunate to have had many volunteers help at the ranch. We couldn't have operated the ranch without their hard work and enthusiasm. Judy and I have hosted events out at the ranch to recognize our *ranch family*—the many volunteers who have made our operation possible. We call our volunteers the ranch family because we do become a family after working together while riding the pastures, doctoring the cattle, and making improvements.

Additionally, we have tried to support our local organizations and to introduce the community to a working ranch. Lunches for guests are set up on tables under the roof of the tack room. For smaller events,

Judy Backus, right, *is overseeing the dessert table at the Christmas party for our ranch family and guests. That year the food was served on the open lot on the west side of the bunkhouse. Our ranch is often used for gatherings and tours. We host foreign guests and show them our operations and the solar installation. We also do fun things like take guests on hayrides, serve lunch, and let youngsters ride a horse around the headquarters. Photo by Jack Carlson, December 2011.*

the meals are prepared in the bunkhouse, but for larger events, a caterer is hired to provide a hot meal.

Every few years, we hold a big event at Christmas that recognizes our ranch family of volunteer cowboys and cowgirls. Musical entertainment is always included, as well as talks and presentations about the Quarter Circle U Ranch history by people with firsthand knowledge of the early years of operation.

Elizabeth Stewart and Jack Carlson at the Christmas party. December 2008.

Chapter 8

Range and Water Improvements at the Quarter Circle U Ranch

The availability and the location of water and cattle rotations are major concerns for Western state ranchers. This often determines and limits the number of cattle and the flexibility of movement of cattle on the ranch.

When we bought the QCU Ranch, there were only natural springs for water sources at the ranch. The historical improvements at these water locations were minimal (just concrete drinkers), and even those had not been properly maintained. Some of these sources of water were developed more than 100 years ago.

The ranch had always been operated mostly with the natural limitations imposed by these mountains and natural waters that existed on the ranch. Fortunately, the location of this very old ranch was such that it was able to operate fairly well. The ranch was mainly located in the foothills of the major mountain ranges that supplied natural water sources. Those mountains also provided natural barriers that could restrict the movement of cattle. Drift fences across mountain passes kept cattle somewhat confined to limited canyons.

Mother cows remained in the same pastures throughout the entire year with certain bulls, which were sometimes rotated to different pastures. The calves were gathered as yearlings and sold by the ranch owner.

However, in the cattle management courses I had studied, they said that a rest-and-rotation program improved the range health and thus provided for range improvement. I studied the range grasses that existed on the ranch, as well as some areas that were inaccessible to cattle or infrequently used by cattle, such as above cliffs or on the tops of mountain ranges. These grasses included some that were very desirable to cattle, which assured me that they had grown there naturally.

However, they were in regions that were inaccessible to cattle and thus were not grazed. I became convinced that those grasses would return with proper rest and rotation management of the cattle. I thus started to plan for where fences could be installed and how the cattle could be rotated in such a way that those desirable grasses would return. The temporary increase in the number of cattle in a pasture would increase the number of cattle that would be forced to those higher elevations to seek those grasses. I had very good advice from the people at the National Resource Conservation Service (NRCS)—and later financial support from them.

After the assessment and recommendations from NRCS, we first set up sampling locations to monitor the progress we made for the proper rest and rotation systems that we implemented. We also set up a schedule for monitoring these sites on an annual basis to identify any progress made on distinct species over a multiple-year time period.

While implementing the major installation of the fencing required for the range rotation and management, we also assessed the requirements for extra water in each pasture due to the increased number of cattle that were required by the shorter use times. This required either the installation of larger storage tanks or for water lines to be piped in from adjacent pastures. Because of the variation of the terrain of the ranch and eventually two ranches, it required a different solution for each of the different pastures. Water development required new spring boxes to be built, storage tanks installed, and extensive new water lines installed.

As a result of these studies, the following actions were taken:

- Two new water wells were drilled at the Northern Ranch.
- Six new solar pumps were installed in wells and springs.
- Twelve water springs and wells had major improvements.
- Thirteen new storage tanks (2,500 to 10,000 gallons) were installed.
- Twenty new water drinkers were installed, with floats to conserve water.
- Ten miles of poly pipe were run to drinkers—some required solar pumps.
- An EQIP grant from NRCS helped buy seven 2,500-gallon storage tanks, nine Powder River drinkers, and 2 miles of 1-inch diameter, 200 psi poly pipe.
- Many new corrals and three new ranch buildings were constructed at the headquarters.
- New and improved handling facilities were made in the corrals to better handle and accommodate artificial insemination, pregnancy testing, weighing, and branding—all with gentle processing for both cows and calves.

The installation of all the above improvements took place over a time span of more than 30 years by ranch employees and various volunteers. Perhaps my family members did not consider themselves as volunteers!

After the acquisition of the Northern Ranch, all the cattle were moved to the north for six months, and Judy and I went with them—to a log cabin we bought there. The Superstition Ranch manager, Howard Horinek, did not move with the cows and thus was available to work on the Superstition Ranch improvements during the summers. These projects usually required at least two people or more. Close friend Jack Carlson was a regular volunteer with Howard. Sometimes they had other volunteers—friends, my family members, and some paid summer students.

The pictures below, mostly taken by Jack Carlson while helping Howard, show some of the projects being installed:

Chuck with his faithful ranch helpers in front of the ranch house. From left: *Chuck Backus, Jack Carlson, Joe Yarina, and Bill Smith, about 2005.*

The 2,500-gallon storage tanks, Powder River brand drinkers, and 1,000-foot coils of poly pipe are roped and staged for the helicopter pickup and delivery to the remote canyon pastures. Photo by Jack Carlson, February 2006.

The Powder River brand drinkers are outfitted with small animal ramps and prefabricated with metal crossbars to hold the float valves. From the left: Chuck Backus, Joe Yarina, Tony Backus, Craig Doyle, Gary Doyle, Jerry Walton, Bill Smith, and Howard Horinek, who is leaning over at the end of the drinker. Photo by Jack Carlson, February 2006.

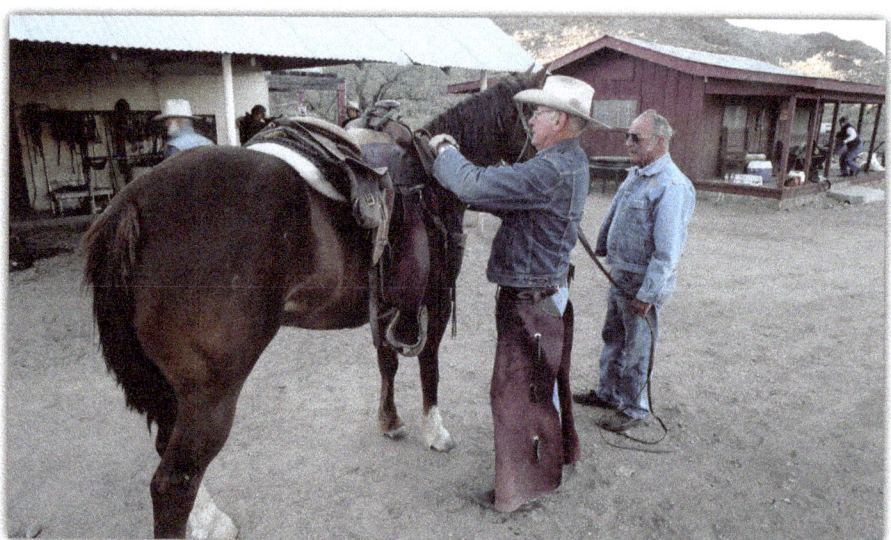

Chuck Backus, left, is checking his gear before the ride to No Name Canyon. Bill Smith, right, is holding the horse. The bunkhouse at the Quarter Circle U Ranch headquarters is in the upper right corner. The tack room is in the upper left corner. Photo by Jack Carlson, February 2006.

Mike Doyle, at the tack room, is a member of the field crew that will ride to the helicopter drop sites. Since 2022, Mike and Amy have been the new owners of the Quarter Circle U Ranch. Photo by Jack Carlson, February 2006.

From the left: *Amy Doyle, Mike Doyle, and Howard Horinek. Mike is part of the field crew, and he will be riding his horse to the remote canyons to help with the helicopter drop. Howard will be riding in the helicopter and navigating for helicopter pilot Ed Boyle. Photo by Jack Carlson, February 2006.*

The ground crew at the ranch headquarters, not shown in this photo, are Bill Smith, Kati Weingartner, and Molly Westgate. They just hooked the load to the helicopter hoist cable. Pilot Ed Boyle is taking the three Powder River drinkers and several 1,000-foot coils of poly pipe east toward the Fraser Pasture drop points, which are four roadless miles away. Ranch manager Howard Horinek is in the helicopter and navigating for Ed. Photo by Jack Carlson, February 2006.

Ranch manager Howard Horinek is prefabricating a metal cover for the house well at the headquarters. He prefabricates the drinkers, solar panels, and other equipment for the spring and well development here at the headquarters before packing the parts by horse to the remote locations. Photo by Jack Carlson, August 2005.

Horses Little Buddy, left, and Fraser, right, are taking a drink at the Coffee Flat Spring trough. Howard is taking a lunch break after packing and distributing salt blocks in the area. Packhorse Jim Dandy, who was carrying the salt blocks, is just out of the photo on the left. When the trough was constructed, Howard prefabricated the metal sides at the headquarters and packed them to the location where they were assembled on a concrete base. Photo by Jack Carlson, December 2005.

View of the open-air front side of the tack room where the horses are saddled and packed with ranch supplies. Packhorse Jim Dandy is loaded with a water pump destined for spring development work at Coffee Flat Spring. Photo by Jack Carlson, 2010.

Packhorse Clem is loaded with coils of extra poly pipe that need to be moved to another location in Lower No Name Canyon. Photo by Jack Carlson, October 2007.

Howard Horinek purchased this small cement mixer for the ranch for the Coffee Flat Spring work that required mixing a lot of concrete. It is powered by a 110-volt motor that is supplied by our portable gas generator. The mixer, generator, and all of the equipment, as usual, had to be packed to the spring by packhorses. On the last part of the project, Bob Beeman and John Amato delivered 29 sacks of concrete by truck via the Elephant Butte Road. For the last leg over an abandoned road, they used an ATV, then transferred the sacks to packhorse Clem for the final leg of the delivery. Photo by Jack Carlson, July 2010.

Ranch work is done on horseback, and supplies are carried by pack animals because the state grazing lease is mostly roadless. Ranch manager Howard Horinek is cleaning out the spring box on Tatum Spring in Whitlow Canyon. Horses from left to right are Fraser, Little Buddy, and packhorse Clem. Cow dog Gus, a Border collie, is on the right. Photo by Jack Carlson, October 2007.

The Upper No Name drinker is constructed of prefabricated metal panels and is mounted on bedrock with some additional concrete. Note that the capacity of the drinker is doubled in size from a square to a rectangle by adding only two more side panels, which was enough incentive for the cowboys to pack the two extra panels to the remote site. Photo by Jack Carlson, May 2010.

A view at the Lower Tule Spring storage tank of one style of poly-pipe unspooler, a.k.a. a Lazy Susan, that positions the 1,000-foot coil of pipe on its side. When the pipe is unwound from the mechanism, it comes off in a straight fashion, not in coils. Photo by Jack Carlson, December 2006.

A view at the Coffee Flat Spring storage tank of a vertical poly-pipe unspooler that was used to pull the pipe down canyon to the drinker. When the pipe is unwound from the unspooler, it comes off in a straight fashion, not in coils. Photo by Jack Carlson, July 2006.

Howard Horinek with packhorse Clem, who is loaded with a shovel, extra poly pipe, and the vertical poly-pipe unspooler on his sawbucks. They are returning from the Coffee Flat Spring storage tank installation project. Clem is one of my best horses, and he can pack almost anything. He doesn't seem to care what he is carrying as long as there is something nearby to eat. Photo by Jack Carlson, July 2006.

The photovoltaic array for the Lower Tule Spring pump. The assembled array was packed to this site using an adapted Decker pack saddle. Howard Horinek is welding two sections of poly pipe together using the special welder that is powered by the 110-volt gasoline generator. The electric cable from the photovoltaic array runs above ground to the Lower Tule Spring 12-volt DC pump at the spring box. The cable will be pulled through the pipe for protection against the elements and little critters that chew on the wires. The welding apparatus and generator were brought in by packhorse. Photo by Jack Carlson, May 2010.

Since it is difficult to pull the electric wire through a long poly pipe, we are separating the pipe into two sections that are connected with a brass coupling. Howard Horinek has the poly pipe secured on the left side of the welder. On the right side, he has the brass coupler, with a factory-installed poly pipe, clamped to the welder. He is getting ready to insert the welder heater between the pipes. Photo by Jack Carlson, May 2010.

Howard Horinek is making adjustments to the water pump enclosure at the Lower Tule Spring. The pump is supplied with 12-volt DC power from the photovoltaic array that is mounted on the nearby hillside. The pump moves water uphill to a 2,500-gallon storage tank on the ridge, which then gravity feeds the water to the Lower No Name Canyon water troughs. Photo by Jack Carlson, October 2007.

A close-up view of the pipe spanning the Lower Tule Spring wash. The pipe is supported with a steel cable that is anchored on both sides of the wash. The pipe is supplying pumped water to the 2,500-gallon storage tank on the ridge overlooking No Name Canyon. Photo by Jack Carlson, May 2010.

View to the southwest into Tule Canyon from the ridge between Bluff Canyon and Tule Canyon. This photo shows the rugged terrain that is typical for all of the ranch pastures. The creek has only seasonal water, so the springs and water catch basins in the canyon are the main source of water for the cattle and wildlife. Photo by Jack Carlson, December 2005.

Jack Carlson is mixing concrete in a feed pan for Howard, who pours the mix into the hardware cloth form at Horse Spring. All of the concrete, tools, and gear have to be packed to the site by horses. Photo by Howard Horinek, December 2006.

At Horse Spring, Howard added concrete to the hardware cloth to anchor the metal collar to the spring box. Howard fabricated the metal collar at the headquarters from scrap metal that I purchased at several auctions. Photo by Jack Carlson, December 2006.

At Mad Cat Spring, Howard is filling the hardware cloth form with concrete. This is the typical layout for constructing a spring box. Later, a metal collar was added to access the inside of the box. Mad Cat Spring got its name from a nearby rock formation that looks like the face of a mad cat. Photo by Jack Carlson, November 2006.

With the prefabricated metal collar in place, Howard is arranging the hardware cloth to define the outside of the Lower Tule Spring box. This spring box will be outfitted with a solar-powered electric pump. The water will be pumped to a nearby ridge and gravity-fed to drinkers in the lower end of No Name Canyon. Photo by Jack Carlson, December 2006.

Metal signs were printed and mounted on the ranch gates and fences to help outdoorsmen know that some gates need to be closed at all times. This gate is in the Southern Pasture, and the inset shows the two versions of the gate sign. Photo by Jack Carlson, March and April, 2006.

Howard is using a Pionjar brand jackhammer to drill a posthole for a new gate that is located in the high cliffs to the southwest of the Upper Corral. The jackhammer was a big improvement in fence building compared to digging or drilling the postholes by hand. We loaded packhorse Tonopah with the jackhammer that was mounted on a special platform on his packsaddle. Photo by Jack Carlson, July 2007.

At the headquarters tack room, packhorse Clem is outfitted with the black post packs that are filled with corner fence posts and braces. The fence post driver is on his sawbucks. If you look closely, you can see the jackhammer loaded on top of Tonopah's packsaddle. He is the last packhorse in line at the tack room. Photo by Jack Carlson, July 2007.

At Tule Divide Corral, Howard would rope a calf, and I would attach a ranch tag to its ear. On one occasion, the mother cow rammed me, resulting in bruises and broken ribs. Howard designed this fence so that I could stand behind the fence while tagging the calf, and that made calf tagging a lot safer for me. Photo by Jack Carlson, October 2009.

A close-up view of the water gap fence anchor in the middle of the Barks Canyon Wash on the northern boundary of the ranch. We are lucky when we can find a solid rock like this to anchor the fence posts. Here the fence is tied with breakaway wires so the flood water will open the fence and let the debris pass. We used a jackhammer to drill the post holes. Photo by Jack Carlson, April 2008.

Packhorse Clem is loaded with a wheelbarrow for spring work in Bluff Canyon. Using a wheelbarrow to move dirt and rocks made working on the springs and water holes a lot easier. Photo by Jack Carlson, June 2005.

A view of the water gap fence on Navaho Wash at the southern boundary of the ranch. The water gap fence is constructed with two wings that are attached in the center of the wash with a breakaway wire. During a flood, the wire will break and allow the two fence wings to open and let the flood debris pass down the wash. The rock baskets add support to the fence posts on both sides of the wash. Photo by Jack Carlson, March 2006.

Chapter 9

Superstition and Northern Ranches, Grazing Leases, and Allotments

Superstition Quarter Circle U Ranch

The Superstition Quarter Circle U Ranch is entirely leased from the Arizona State Land Department. For a ranch with a defined number of acres or sections of land, the Arizona State Trust land grazing lease specifies only the maximum number of animal-unit-months (AUMs)—the number of cattle multiplied by the months of grazing—that a rancher on Arizona State Trust lands is allowed to graze.

For example, on our Superstition ranch, we can graze 250 head year-round, or we can graze 500 head for six months of the year. This allows the Arizona State Trust lands to better support grazing in the parts of the state that are more arid.

In most of central Arizona, we receive more rain during the summer months, and it supports many seasonal grasses. Those seasonal grasses grow in the summer months only, while the annual grasses grow throughout the year. Many ranchers in these arid areas have a base head count of cows that are on the land for 12 months, and the ranchers lease out their pastures for use by feedlots during the summers—as long as they stay within their year-round AUM units.

The Superstition QCU Ranch leases more than 22 square miles of Arizona State Trust land and was rated for about 215 mother cows—year-round. The calves are not counted until they are near yearlings. In most of my ownership years of the ranch, it has been operated as a year-round, cow-calf operation.

In 2000, at the age of 63, I started to plan to retire from ASU to become a full-time cattle rancher. I thought that it would then be appropriate for me to own two ranches—one in the area of the present Superstition QCU Ranch and one in northern Arizona—in the high country at more than 7,000 feet. That way, I could double the number of cows at the Superstition base ranch for six months and then move the cows to a summer ranch for the six months of the summer. If Judy and I could buy a home near there, either on the new ranch or off the ranch, we would have a more comfortable life, both summer and winter. Thus, we started looking for ranches for sale in the Pinetop-Lakeside area that had a lease for at least 450 cows for the six months of summer. Almost all of the undeveloped land surrounding the Pinetop-Lakeside area was owned by the U.S. Forest Service. Thus we had to find a USFS ranch that met those requirements.

Northern Ranch

At the summer meeting of the Arizona Cattle Growers Association (ACGA), I asked the president, Jake Flake, from Snowflake, if he knew of any ranches in that area of the state that could accommodate at least 450 cows for the six months of summer. He looked surprised and said, "Yes, we have one that we would like to sell." It turned out that he and his brother owned several ranches in the area and wanted to split them up in order for the two of them to retire. Further discussions and a visit led to our purchase of his ranch.

This new purchase occurred quickly and long before I had planned to own a second ranch. I had planned to retire in the summer of 2003 (I actually retired in 2004); thus, I was expecting to own a northern ranch to which I could move cattle that summer of 2003. The addition of another ranch obviously required many changes to the operation of the original Superstition QCU Ranch.

The newly acquired ranch was often referred to as either the Northern Ranch or as QCU North. The rest of this book discusses the joint operation of the two ranches that we operated as one ranch.

Improvements in the Pastures

The ranch we bought from the Flakes was actually two U.S. Forest Service allotments—one allotment was located mostly south of Porter Mountain Road and the other was an adjacent allotment that ran north of US 60—east of Show Low—for a total area of 40 square miles. These allotments were at about 7,000 feet of elevation, and the terms of the grazing allotment were for use only during the six months of summer, which was all we wanted. Of course, this required that we truck, twice per year, the eventual number of more than 400 cattle and their calves between the two ranches.

This new Northern Ranch had been neglected for many years and really needed the elimination of young trees and underbrush. I applied for all kinds of grants to do this clearing and water development. I was successful and received many grants, which included grants to drill 2 new wells and add a distribution system to send water to new regions of the allotment. These improvements allowed for the proper movement of cattle, which would improve the grass in the pastures.

Three generations of ranchers/cowboys. At the left is Chuck; in the middle is Mike Doyle, who, with his wife, Amy (our daughter), bought our Superstition Ranch in 2022; and at the right is their son, Sean Doyle. This picture was taken at the Northern Ranch.

Small Caterpillar bulldozers were used to uproot trees and stack them into windrows. Photo by Chuck Backus, June 2006.

Cassy Murph, our Northern Ranch cowboy, riding between the trees that had been uprooted. Photo by Chuck Backus, June 2006.

A view of one pasture shows the area where the trees were removed. The few trees that were left are for the benefit of the cattle. Photo by Chuck Backus, June 2006.

The cows and bulls were allowed to return to the pastures after they had recovered. Photo by Chuck Backus, June 2006.

At some point, for our pasture rotation plan, the cattle had to cross US 60 to the pastures on the north side of the ranch. The Arizona Highway Patrol had to assist in this crossing. Photo by Chuck Backus, July 2010.

We drilled 2 new wells on the Northern Ranch pastures, converted all the wells on the ranch to solar-generated pumps, and expanded the capacity of the water storage tanks. Photo by Chuck Backus, June 2007.

Headquarters for the Northern Ranch

Judy and I bought a large, 5,000-square-foot house on three acres in Lakeside, Arizona, that could accommodate many family members and temporary ranch helpers. We could also accommodate many horses. Plus, we built a large barn with an adjacent corral. We bought this property at about the time of my retirement from ASU so Judy and I could move completely up to the Lakeside house with the cattle each summer. I was kept very busy improving both the Lakeside house—the property had sat empty and unkept for more than four years—and the Northern Ranch, which had also been neglected for years. However, owning a large home at 7,000 feet of elevation with cool Arizona summers made the location attractive to many family members and friends who visited and helped at the ranch.

The Superstition ranch manager did not move to the Northern Ranch with the cattle for the summer, thus making him available to work on major water improvements at the Superstition ranch. My close friend Jack Carlson was a tremendous help in water development during all those years, during the summers and at other times. The water developments that Howard and Jack built are discussed in Chapter 8.

I located my ranch office in one room of the headquarters house.

This is a drone view of our large, 5,000-plus-square-foot, two-story log cabin that we bought for the headquarters for the Upper Ranch, a.k.a. Northern Ranch.

The front room was very large and had an open ceiling, a fireplace, and a kitchen.

We also built a large storage building that housed a garage, a barn, and an apartment. The barn could hold a full trailer of hay, and the apartment for the cowboys was complete with a kitchen and bathroom.

The ranch manager's quarters are inside the east end of the barn, and the apartment is equipped with a kitchen, bedroom, and bathroom.

The main entrance is on the north side porch of the headquarters house.

Chapter 10

Headquarters Handling Facilities

Headquarters Corral Design

By 1980, with all of the fencing work that was going on in the first few years of owning the Superstition QCU headquarters ranch, I could not concentrate on much else. It was obvious that the handling facilities at the headquarters were inadequate for a modern ranch.

When we bought the ranch in 1977, the corrals, as shown in figures 10.1 and 10.2, consisted of the stone barn, built by Jim Bark in 1891, and a board-fence holding pen to the south of the barn. There was a board-lined loading ramp to load and unload cattle from a truck. A large square-wire metal fence enclosed the west side of the barn where cattle could be gathered. The facilities were capable of only collecting and shipping cattle. Thus, I started to evaluate what needed to be done.

First, the adult cattle have to be managed and processed in a way that is safe and easy for both the animals and the people involved. There also needed to be a way to safely separate the bulls from the cows and also the cows from the calves. The calves needed to be worked (branded, castrated, and tagged) in a humane way with as little pain as possible.

Figure 10.1. View of the headquarters facilities showing the stone barn and wire-fence corral for the horses about the time that we bought the ranch. Note the standalone squeeze chute on the far right. Photo by Tom Kollenborn, circa 1977 or earlier. Courtesy of Keith Ferland and the Kollenborn Collection.

Figure 10.2. View of the headquarters facilities—stone barn, windmill, and a corral—about the time that we bought the ranch. Photo by Tom Kollenborn, circa 1977 or earlier. Courtesy of Keith Ferland and the Kollenborn Collection.

The Installation

The new fence posts for the cattle chutes and corrals were made from 6-inch-diameter steel posts with welded-on 7/8-inch horizontal steel sucker rods for strength. The steel material avoided wood splinters for both cattle and humans. All fences were a minimum of 6 feet high. An 8-foot-wide lane was constructed for sorting, with 5 corrals available to sort cattle into. All of the corrals have water drinkers.

A view to the northeast of the Quarter Circle U Ranch headquarters in 2007. From left, the white-colored building was the ranch manager's residence that was removed in 2018; next is the solar panel array, the bunkhouse, the tack room, the working corral area, the hay barn, and far to the right is the old stone barn and windmill. The dry waterway below the headquarters is Navajo Wash, which originates near the Upper Corral and Well. Photo by Jack Carlson, October 2007.

Another set of smaller corrals was built in a row, leading toward the main set of the working area corrals. This smaller set of holding pens allows the cows to be separated into groups that may require different types of processing or semen from different bulls.

One large section of the working area corrals is covered with tin roofing that provides shade and a dry space to process the cows. The shade eliminates shadows and reflections, which often frighten the cows and impede the movement of cattle. The presence of coats or other objects hanging on the sides of these lanes in the working area is always avoided. The shade not only protects workers from adverse weather conditions, but it also protects vaccines from exposure to sunlight and heating.

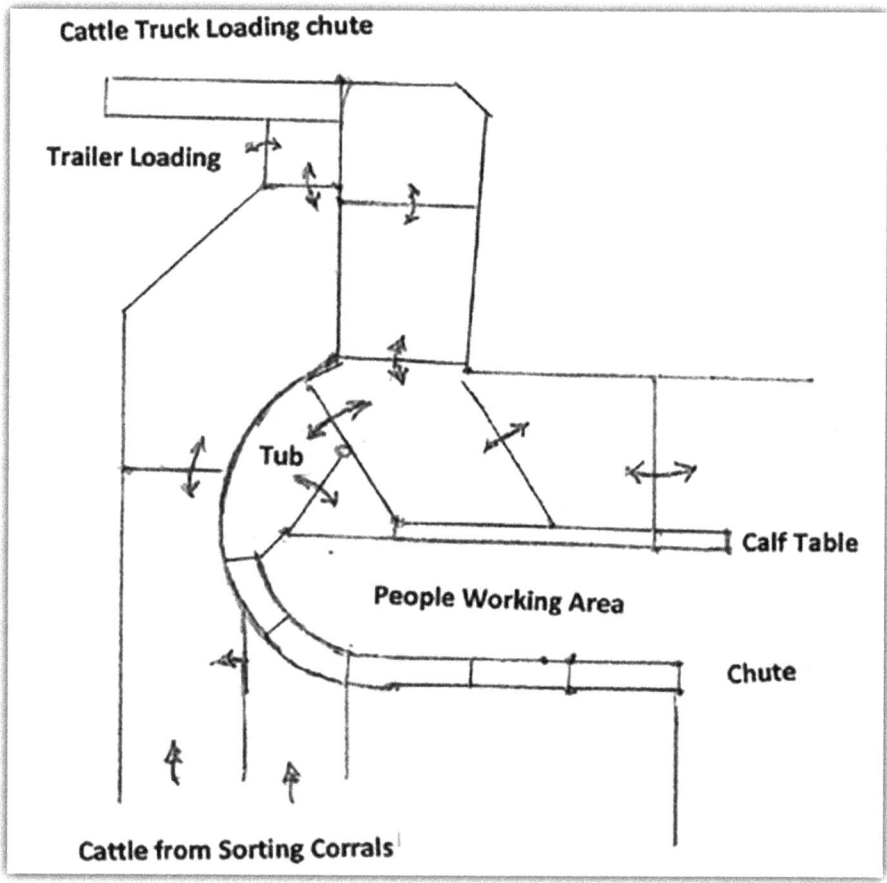

This rough schematic shows the layout of the working area corrals. The cattle come from the sorting corrals (bottom), go up past the loading chutes (top), and then down toward the Tub (center). The Tub gate either directs the cows into the single-animal queuing stalls leading toward the adult squeeze chute or directs the calves toward the calf queuing lane leading up to the calf table. The people working area is in the middle to allow them to process either adult cattle or calves. Both lanes and the people working area are shaded for more effective processing of the cattle. Sketch by Chuck Backus, 2017.

Sorting cattle at the headquarters corrals. I am sitting on the railing, keeping the tally. I had many friends and relatives that helped with the cattle processing. The helpers are looking toward the lane between the corral pens where the cattle are staged. We add cattle to this lane and then sort them into the desired pen according to their processing needs. Photo by Jack Carlson, April 2005.

An overall view of the working corrals. Note the cover over the area where the people work and the animals are processed. The solid wall in the foreground is the outside wall of the Tub. Photo by Chuck Backus, 2017.

Operation of the Processing Facilities

The cattle enter the working area by first entering a circular *Tub*, which is a solid steel gate that swings from a central pivot post (see figure 10.3). By swinging this gate behind the animals, the cattle can be directed into a single lane that is 27 inches wide (see figure 10.4). The single lane leads to the queuing stalls, where the cattle can be held while waiting to enter the squeeze chute.

Another view of the corrals looking northeast with the branding area on the left and the hay barn in the center. Photo by Jack Carlson, December 2011.

The queuing stalls are separated by sliding wooden doors, allowing the animals to be advanced one at a time (see figures 10.5, 10.6, and 10.7). There are 4 queuing spaces in our design. This set of queuing spaces allows one person to bring up cattle from the corrals and fill the spaces while other people are continuously processing the cattle through the weight scale and the squeeze chute. Sometimes the cattle may be vaccinated or otherwise processed before entering the squeeze chute.

The queuing space directly before entering the squeeze chute has an access door from the people working space that lets a helper enter the queuing space (see figure 10.8). This allows a person to step in behind an animal standing in the squeeze chute for certain processing, such as pregnancy testing, artificial insemination (AI-ing) of cows, physical examination of bulls, etc.

Figure 10.3. The original circular Tub that cattle come into when entering the working area corrals. The solid steel gate is swung around to direct the cattle into the narrow entrance of the single-animal queuing lane, which is in the upper left corner of the photo. Photo by Chuck Backus, 2009.

Figure 10.4. A modification to the original Tub was made by installing an additional gate with wheels on its end and with a spring that holds the added gate against the original gate to more effectively direct the cattle toward the single lane. The solid sides keep the cows from being distracted. Photo by Chuck Backus, 2017.

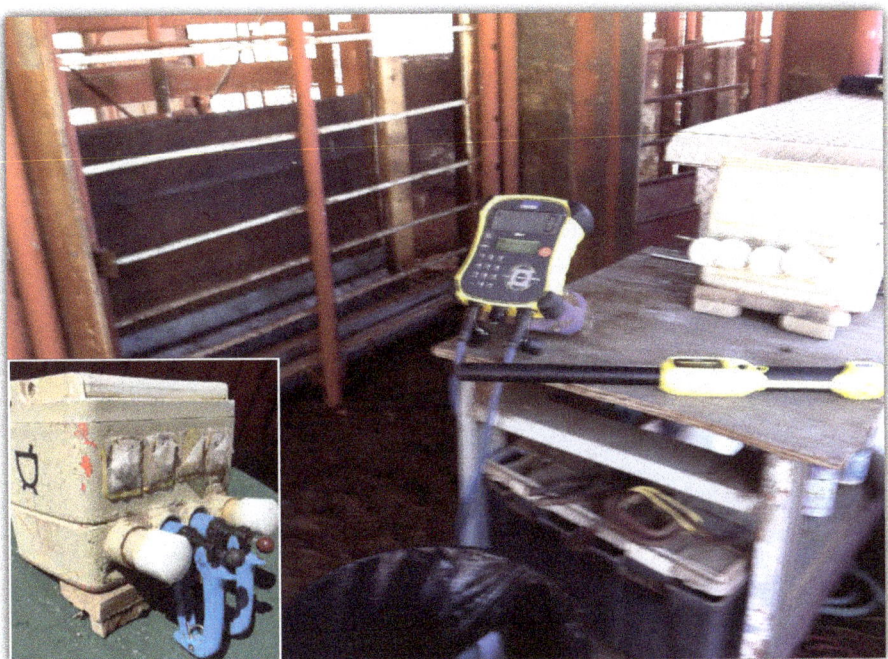

Figure 10.5. The queuing stalls are divided by wooden panels that move horizontally to keep the cattle separated. The central stall, shown here, has an inline weighing scale with an aluminum platform for electronically weighing cattle. The long hand wand on the table is waved by the cow's head to scan the electronic tags. Then the scale recorder/computer on the table displays the cow's EID tag number, the ranch tag number, and the weight of the cow. The ice chest, behind the wand and also enlarged in the inset, is a homemade holder for the syringes used for vaccinations. Between usages, the syringes are placed in the front opening of the tubes going into the side of the cooler. The cooler is filled with ice and bottles of vaccine. Photo by Chuck Backus, 2016.

Figure 10.6. A view of the queuing stalls. Stall 3, where a cow is standing on the left, is where we eventually installed the electronic scale. The cow on the right is in stall 2. The ranch crew is assisting with AI-ing the cow in the squeeze chute. Photo by Jack Carlson, April 2006.

Figure 10.7. This calf is being weighed on the new electronic scale. We sorted the calves into groups of 10 in 7 pens, and then Jack brought the groups up to the processing area for weighing and vaccinations by Howard Horinek and Katie Weingartner. On that day, we processed 87 calves before noon while we waited for the veterinarian Larry Lunt to come out to preg test the adult cows. Photo by Jack Carlson, October 2006.

Figure 10.8. The white gate to the right of the red and grey squeeze chute allows access to the queuing space. Veterinarian Larry Lunt, left, is opening the wooden sliding door to advance a cow through the queuing stalls. The stall to the right of Larry has the inline weight scale. When Larry performs the pregnancy evaluation, he steps in the stall through the white gate and stands behind the cow that is in the squeeze chute. Other helpers, from left to right, are Merlin Yeager, sitting in the chair, who is preparing a vaccine shot; an unidentified person, and Molly Westgate at the weight scale display. In the foreground on the barrel are some of the vaccines and applicators. Photo by Jack Carlson, October 2007.

The second queuing space back from the squeeze chute contains a single-animal electronic weighing scale. This inline weighing setup allows the weighing of the animal each time it is processed through the squeeze chute (see figure 10.9). An animal's weight change through its lifetime can be used when evaluating the animal at a later date.

Since all animals have an electronic ear tag, they can be identified, and their history can be retrieved by using an electronic identification (EID) reader. The cordless Bluetooth reader is passed over the animal's head, and the animal's identification is transmitted to the scale monitor and computer. The scale monitor shows the electronic ear tag number of the animal, the plastic ranch ear tag number of the animal, and the current weight of the animal. A push of a button electronically stores this data for later review and/or transfer to a ranch computer. The electronic weight scale and EID reader greatly reduce the processing time of the animal and improve recording accuracy. For a photograph of the ear tag, see Chapter 12, "Calf Processing in the Spring," in section "Electronic Identification (EID) Tags."

After weighing, the animal proceeds from the scale to the next queuing space and then into the squeeze chute (see figure 10.10). Before the animal enters the squeeze chute, the automatic head catch is set inward so that when the animal puts its head through it, the shoulders of the animal close the head gate and lock the animal in place. In contrast, older squeeze chute designs have a long handle on the headgate. The operator was required to quickly time the closure of the headgate with the position of the animal coming through the squeeze chute—to catch the animal's neck just at the right time. Being off slightly with the timing of the handle results in slamming the gate on the animal's head or letting the animal get away.

Figure 10.9. The recording bench in the people working area shows the scale recorder/computer, the spiral-bound notebook for recording data, and the branding iron. Dr. Larry Lunt and Chuck Backus are in the background. Photo by Jack Carlson, 2009.

The squeeze chute has high vertical doors on either side of the animal's neck for proper placement of the vaccinations (see figures

10.11 and 10.12). The two sides of the squeeze chute are vertical, and both move horizontally to *squeeze* the animal into position. This means that the space for their feet is as wide as the width of the animal. The wide, flat floor makes the animal feel secure while standing in the chute. Several smaller side gates allow access to the sides of the animal and to their feet for any necessary treatments.

Figure 10.10. The squeeze chute. The neck-access doors, on both sides, allow vaccinations following the beef quality assurance (BQA) recommendations. The two sides of the chute move horizontally, allowing the cow to have secure footing. The head gate allows the cow to automatically secure herself by putting her head through the opening with her shoulders pushing the gate closed. The vertical-opening back gate allows access from the back. Photo by Chuck Backus, 2016.

An aluminum vertically operated back gate can be opened and closed with a rope to contain the animal or for needed processing from the rear. The outlet from the squeeze chute is into a small holding corral to allow for later movement to another desired location.

Figure 10.11. Chuck Backus, right, is preparing a CIDR for AI-ing as Howard Horinek, left, assists at the squeeze chute. Photo by Jack Carlson, April 2006.

The processing facilities for calves are on the opposite side from the adult cattle lane of the people working space, but still under the same shade cover. The calves are moved up to the Tub location but go by the backside of the Tub gate into a calf-sized single-lane chute. A gate, which serves as a side of the calf lane, is opened and allows 4 or 5 calves to enter the lane. The lane gate is closed against the calves with the calves all facing the same way (see figure 10.13). The calves are moved up this single-wide lane, which has the *calf table* at the end (see figure 10.14).

When a calf enters the table, the calf's head is caught. One side of the table moves inward, allowing the calf to be squeezed against the tabletop. Then the table, with the calf squeezed against the tabletop, is rotated 90 degrees, putting the calf into a horizontal tabletop position—thus the name, calf table. The calf table is 27 inches wide. The calf table constrains the calf in such a way that it can't move around or hurt itself or the operator. The calf table is used for the safe processing of the calves, such as electronic ear tagging, vaccinating, ear marking, branding, castrating, and any needed doctoring (see figures 10.15, 10.16, 10.17, and 10.18). For more about freeze branding, see Chapter 12, "Calf Processing in the Spring," in the section "Freeze Branding."

Figure 10.12. Merlin Yeager is getting ready to open the head gate to release the cow from the squeeze chute. The cow had just been pregnancy checked by veterinarian Larry Lunt. Photo by Jack Carlson, October 2008.

Figure 10.13. The cattle are in the queuing lane with three calves waiting to be processed. The fence on the back side of the calves is also a gate that is hinged in front of the first calf. To load the lane, that gate is opened to be parallel with the back of the Tub gate—seen beyond the backs of the calves. As the calves enter between the two gates, the calf lane gate is closed with all the calves headed toward the calf table end of the lane. Photo by Chuck Backus, 2016.

Figure 10.14. The calf table. Individual calves enter from the right and have their heads caught by the head catch on the left end. The open-barred side is pulled up to squeeze the calf toward the solid side, and then the table is rotated 90 degrees to put the calf in a secured, horizontal position. This antique table was given to Chuck by his mentor, Lee Thompson. The open-barred side is vertically hinged at the back, allowing the calf to exit when the table is rotated back to the vertical position after the processing is completed. Photo by Chuck Backus, 2017.

Figure 10.15. A calf is secured on the calf table in the horizontal position while Merlin Yeager sprays 99% alcohol on the shaved hide of the calf in preparation for the freeze branding. A low percentage of water in the alcohol is used because water will freeze and interfere with the branding. The bars on the calf table can be opened and laid back to allow complete access to the side of the calf. Photo by Jack Carlson, October 2007.

Figure 10.16. A calf is on the calf table in a horizontal position with one of the side bars opened. Dean Harris, left, is pressing the freeze branding iron on the side of the calf with his right hand and looking at his watch on his left hand to time the brand. The cold, minus 344°F branding iron (brass in this case) is held against the hide for about 40 seconds. Eddie Christopher, right, is preparing to castrate the calf as soon as Dean completes the branding. Photo by Chuck Backus, 2015.

Figure 10.17. A view of the empty calf table in the working horizontal position. Photo by Jack Carlson, October 2006.

Figure 10.18. As a cow matures, the freeze brand grows in size. The white freeze brand on cow 413 is easy to see on her contrasting black hide. After pregnancy evaluation, veterinarian Larry Lunt marked her back with an orange grease marker with the number 5 to indicate the number of months that she is pregnant. Photo by Jack Carlson, November 2005.

Chapter 11

Cattle Operations Schedule

I have always considered myself a grass farmer who utilizes cattle to improve and harvest the grass in order to provide a healthy protein source for humans. I have set three general goals to successfully do this:

1. Landscape goal: to improve the diversity and health of native forage on the ranch.

2. Herd goal: to increase the number, weight, and quality of calves raised on the ranch.

3. Family and community goals: to provide a successful ranch experience for my family and friends and also to provide research, education, and service to the greater community.

To accomplish these goals, I have used *modern range management*, which can be defined as the use of plant, animal, and range management sciences to achieve a desirable, productive, and sustainable landscape that provides food and fiber for society. The operation and practices I have implemented to create such a ranch have now spanned more than 50 years. The cattle operations can best be described by breaking them into several stages during the year.

October–Processing Cattle from Summer Pasture

The cattle are gathered from the Northern Ranch summer pasture during the last two weeks of October. Then the cattle are shipped, one truck per day, during the workweek days to the Superstition Ranch. The cows and calves are kept in separate compartments on the trucks. On arrival at the Superstition Ranch, the cows are put in the main corral for feeding until the Saturday processing. The calves are processed as they come off the truck and put in separate corrals from their mothers. This is the official weaning time for the calves, although most have been weaned by their mothers over the summer. Processing of the cows and yearling heifers is done on the last two Saturdays of October when the veterinarian comes out to examine and pregnancy test (preg test) the animals.

Dean Harris is seen here driving his truck into the Quarter Circle U Ranch headquarters with cattle from the Northern Ranch. Photo by Jack Carlson, October 2008.

Unloading cattle from the Northern Ranch directly from Dean's truck into the corral area. The corral alley has a built-in loading ramp that the truck can back up to. Photo by Jack Carlson, October 2008.

The corral fences and gates are arranged so the cattle can be unloaded from the truck and directed down to the sorting pens. This section of fencing is also used to bring cattle into the people working area for processing. Photo by Jack Carlson, October 2007.

October–Calf Processing

Right after the calves are unloaded from the truck, they are run through the corral working lanes and given the following treatments:

- Ultra Bac 8—an 8-way Clostridial Booster (black leg, etc.)
- A Pour-On—for various internal and external parasites
- Bovi Shield Gold One Shot—a booster MLV with Pasteurella
- Inforce 3—a nasal spray, MLV for instant protection
- Trich-Guard V5L—for heifer calves—(first dose)
- Vitamins A, D, and E

The total cost of vaccines for the life of a calf is about $10 per calf.

During processing, all the calves are weighed on the electronic scale. That way, we can calculate the average daily gain (ADG) from the time they were branded the previous spring. The ADG as well as their physical appearance will determine which calves are poor performers and thus likely candidates for selling directly. The other steers and heifers that are not candidates for selling are kept separate from the cows and *backgrounded* for six weeks.

Backgrounding is the term used for preparing calves for the feedlot environment or for heifers that are to be kept as replacement cows at our ranch. It is better to background the calves at the home ranch. Calves seem to do much better in familiar locations. This also gives the rancher a longer time to select the calves that may be best for replacement heifers. However, the main issue in replacement heifer selection should be in the detailed study of the genetics and performance of her mother and father, if it is known. Genetics and performance are better indicators than physical appearance of her potential value to the herd.

A view to the northeast of the main corral where the cows are feeding at the hay barn feed manger. When the calves are initially fed here at the home ranch, they adapt much better to the feedlot environment and stay healthy. Photo by Jack Carlson, April 2006.

A view to the southwest from the hay barn to the main corral feed manger. With the hay barn adjacent to the manger, feeding the cows is much easier than hauling hay to other locations. Photo by Jack Carlson, November 2006.

The calves are fed relatively low-quality/priced hay during this time. This backgrounding at the home ranch where they were born creates less stress and is highly recommended by feedlots. It allows the calves to become completely weaned and become used to eating from a feed trough and drinking from a water drinker. Backgrounding allows the calves to grow out their bodies before transitioning to a higher energy feed at the feedlot. It certainly prepares them for the feedlot and reduces feedlot sickness and deaths.

The backgrounding time also allows for visually evaluating and selecting replacement heifers. DNA samples are taken on most of the heifer calves at branding, and the better heifers have their samples sent into either Gene Seek or GMX Advantage for their analysis and ranking. The DNA results are combined with the physical examinations of their appearance and temperament to determine which ones to keep as replacement cows. The past performance of their respective mothers in the herd, as well as the genetic data of their sires, also goes into the selection process. The selected replacement heifers are given a dose of Multi-Min 90 and a booster Trich-Guard shot, then later turned into a special heifer pasture. During the backgrounding period, we frequently walk through and talk to the calves to get them accustomed to people on foot. This makes them more comfortable around humans.

October–Adult Cow Processing

For five weekdays, the cows are trucked to the Superstition Ranch from the Northern Ranch summer pastures and put into corrals so they will be ready for processing on the following Saturday. Veterinarian Dr. Larry Lunt comes out from Gilbert to examine about 200 head on the first Saturday. He pregnancy checks the cows and calls out the number of months pregnant, to within a half of a month. The number is written in larger figures on the back of the cow with a pink or orange grease marker, with a plus sign to indicate a half month. The cows that are not pregnant, *open cows*, are marked with a zero and then separated from the herd. They will go to the sale barn unless there are unusual circumstances that might justify keeping them. The cows are also physically inspected for any conditions that need to be treated.

After his pregnancy evaluation, veterinarian Larry Lunt marked these cows with a pink grease marker to indicate the number of months that they were pregnant. The cows in the foreground are marked 6+, 5+, and 5. The plus mark indicates half a month. Photo by Jack Carlson, October 2009.

The pregnant cows are then processed with treatments given below:

- Ultra Bac 8—an 8-way Clostridial—a booster shot
- Bovi Shield Gold 5L5—a MLV booster shot with a Lepto additive
- Vibrin—protect against Vibriosis

- Trich Guard Booster
- Pour-On—parasite protection
- Vitamin A, D, and E

On the next day, Sunday, the marked cows are separated and driven to one of two pastures. The cows marked as 6+ through 7+ months pregnant go to one pasture, and those marked as 4+ through 6 months pregnant go to the second pasture. With the processing complete for the cattle that were brought to the ranch during the first week, the crew gets ready for the second week to receive animals on Monday.

October–Bulls Put in Separate Pasture

The bulls are taken from their separated pasture at the Northern Ranch to a special bull pasture near the Superstition Ranch. It is a 1,000-acre pasture with no cattle on any of the adjacent sides. If feed is low in the pasture, sometimes a protein supplement tub is put out. The bulls are brought to the headquarters in February or March for testing and processing before being turned out with the cows—after the cows are artificially inseminated (AI-ed) in April.

November and December–Calves Sent to the Feedyard

The calves are backgrounded at the headquarters from the first of November until the middle of December. The replacement heifers are selected and eventually put into the heifer pasture.

All the steers and the lower end of the heifers are kept at the headquarters corrals to either be sent to the Cattlemen's Choice Feedyard in Oklahoma (with retained ownership by me) or sold at the local sale barn. For this group of animals, we usually have two truckloads of calves, about 150 calves, that we send to the feedyard, and the rest we sell locally. Some percentage of the calves (25% to 50%) are sold to the owner of the feedyard.

The feedyard in Gage, Oklahoma, was selected for feeding our calves since it is a small, custom feedlot that was given the CAB Small Feedlot of the Year award the year before we chose them. They feed only 10,000 cattle as opposed to larger feedyards with up to 100,000 head. They also specialize in what are called *program cattle*. That means they feed cattle for special programs (Natural, NHTC, Age and Source Certified, GAP, etc.) that qualify for premium prices at the packing plants. Our calves are all certified by a third party, IMI, for these programs.

Feedyard owner Dale Moore feeds appropriately to keep the calves certified for these programs and markets the finished cattle to the various packing plants to ensure receiving the highest price. He also buys hedged future contracts for these cattle to guard against fluctuations in the market price.

Detailed carcass data is reported back to us on every calf harvested. With that data, we can evaluate the mother of each calf. If a calf is reported to have graded anything less than Choice, the mother of that calf is sold. Mothers that produce prime-grade calves have a notation in their record to be retained in the herd in case there is a questionable condition with her.

December to February–Perennial and Annual Grass Natural Feed

The perennial grasses on the Superstition Ranch pastures are warm-season growers that grow during the summer, while the cattle are at the Northern Ranch. These grasses go to seed and become dormant by the time the cows are put into the Superstition Ranch pastures in November. At that time, the rains and cool weather have not yet started the annual grasses. The natural feed in the pastures is low in protein at that time, and, therefore, protein blocks are put out in December, January, and February. In the life cycle of the cows, these three months represent the month before calving, the month that most calves are born, and the month after calving. These three months are the time when the cows need a protein supplement the most.

View to the west into Whitlow Canyon from Cat Face Hill. Willow trees in the bed of Whitlow Canyon can barely be seen at Bee Spring. The seep is at the lower right. Photo by Jack Carlson, December 2005.

December–Tagging the Calves

By Christmas, the cows start calving in the pasture with cows marked as 6+ through 7+ months pregnant. Throughout the calving season, the ranch manager and I ride to look for new calves at least every other day and every day when the AI-ed cows are due. Each new calf is roped, and a small yellow ear tag, starting with number one, is put in the left ear, and data is recorded, such as the calf's tag number, the calf's estimated date of birth (DOB), sex, color, and the tag number of the mother.

No Name Canyon. A view from No Name Spring and the cement trough to a cow above the cliffs on the nearby horizon. There is a cow bed ground beyond the cliffs on that mesa that the cows access by a gentler slope. Photo by Jack Carlson, December 2005.

A view of the metal-fabricated water trough in the upper part of No Name Canyon. One of the 2,500-gallon storage tanks can be seen up on the hill in the center of the photo. Horse Tonopah is in the photo. Photo by Chuck Backus, December 2010.

There is an effort to tag the new calves within a couple of days after birth for three reasons: (1) the DOB helps determine if the calf is from a unique AI session, therefore identifying its sire; (2) it is a way of tracking how many calves are lost between birth and branding from losses, usually due to mountain lions; and (3) the new calves are much easier to rope than calves that are a week or two old. When the calves are about two months of age, the cows and calves are gathered from their pastures and brought to the headquarters to brand the calves and to AI the cows for the next season.

February–Bull Testing

In February, the bulls are brought from the bull pasture to the headquarters for processing and testing. Dr. Lunt performs a physical exam, including palpation, to determine any potential physical problems. He then measures the scrotum circumference and takes a sperm sample for the measurement of the sperm motility and number and type of defects. A sample is taken from the bull's sheath for sending to a lab for PCR testing for trichomoniasis (Trich). These tests are done every year on every bull.

A DNA sample is taken on bulls that are not yet in the data bank. When heifer DNA samples are sent in to be analyzed, having the bull DNA on record allows the sire of each heifer calf to be identified. Those tests are a bull test as much as a heifer test, in that they count the number of heifers sired by any individual bull. That is a way to determine which bulls are producing calves and which are not. After all the testing and processing, the bulls are put into the Lower Pasture to wait for the cows to be AI-ed, and then the cows are turned in with them.

The processing of the bulls includes giving their annual vaccinations as follows:

- Bovi Shield Gold 5L5—with Lepto
- Vibrin—protect against Vibriosis

Howard Horinek is checking the new water trough that he fabricated and installed in the Bull Pasture. There are no springs or wells on the two-section pasture, so he connected the pipes to the metered commercial water system in the nearby housing development. Photo by Jack Carlson, July 2009.

Veterinarian Larry Lunt is performing fertility exams on the bulls and evaluating six characteristics of each bull's fertility. He is viewing a bull sperm sample on a glass slide on the microscope and making a tally of the abnormal sperm on the counter with his left hand. He checked about 16 bulls during this session and found one bull that was infertile. We planned to send that infertile bull to the sale barn, but a cattle buyer stopped by to purchase three young bulls. She also asked if we had any animals she could use for hamburger, so we told her about the infertile bull and she purchased it. Photo by Jack Carlson, February 2005.

Veterinarian Larry Lunt, right, is performing a rectal exam of the bull. After a complete examination of each bull, a decision is made to proceed with a vaccination and return the bull to the herd, to doctor the bull to cure its ailments, or to sell the infertile bull to the sale barn. Howard Horinek, left, is assisting. This is a good view of the gate that provides access to the back end of the squeeze chute. Photo by Jack Carlson, February 2005.

- Ultra Bac 8—an 8-way Clostridial—a booster shot
- Vitamins A, D, and E
- Pour-On—parasite protection
- Sometimes, Multi Min 90 V

This bull that failed the fertility test is being weighed before being sold. The bull had not eaten in a while and only weighed 1,780 pounds. This scale was later replaced with an inline electronic scale in one of the stalls leading to the squeeze chute. Photo by Jack Carlson, February 2005.

February and March–Artificially Inseminating Cows and Heifers

In late February, the cows and new calves are first brought in from the pasture that has the oldest calves. The cows are separated from the calves for processing. The first step of the 10-day artificial insemination (AI) protocol is to synchronize all the cows so that they come into heat within the same time interval. The group of cows will then be artificially inseminated at the same time. The procedure used at the

QCU Ranch is called *Fixed-Time AI* and is described in Chapter 13, "Cattle Operations: Artificial Insemination."

April–Bull Selection

Bull selection for live bulls and bull semen is discussed in Chapter 13, "Cattle Operations: Artificial Insemination."

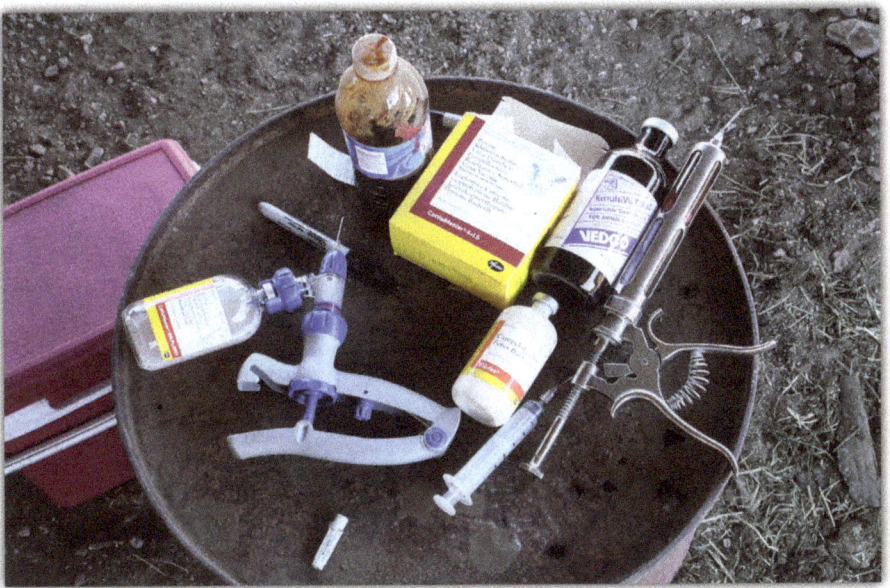

Several empty vaccine bottles and syringes used in processing the cows and calves are shown here. Photo by Jack Carlson, November 2006.

Packhorses Jim Dandy, left, and Clem, right, are each loaded with about 200 pounds of protein blocks, six blocks at 33 pounds each, that are being taken to the Tule and No Name Pastures. Next to Clem's front legs is a stack of salt blocks ready to be packed to remote locations on another day. The following day, more protein blocks were packed down Navajo Wash to the Southern Pasture and dropped at Powder, Horse, White Rock, Saddle, and Castro Springs and a few other locations. These blocks supplement the natural vegetation in the pastures. Photo by Jack Carlson, February 2006.

Packhorse Jim Dandy at the headquarters is loaded with 150 pounds of salt. The 50-pound block on the top platform is for the Tule Divide Corral. Two other blocks were split in half and loaded on the wooden panniers. They were scored in quarters so we could leave a smaller amount if needed. The flat platform of the Decker packsaddle allows us to balance the load by placing the odd weight on top. The salt blocks are being taken to Bluff and Tule Canyons. Photo by Jack Carlson, December 2005.

Chapter 12

CALF PROCESSING IN THE SPRING

In the spring, during the 2+ days that the cows are kept in the corral before being AI-ed, the calves are taken off their mothers, branded, and processed. These processes include freeze branding, weighing, EID tagging, bull calf castrating, DNA sampling for heifer calves, and vaccinating.

After processing, the calves are weighed on the electronic scale, and the ranch tag and EID tag numbers are verified and recorded along with the calf's weight. The calves are then returned to their mothers.

Spring is a very busy time on the ranch, and it is extremely important to the life cycle of the cattle operation.

Freeze Banding

Branding has evolved over the years at the QCU, and we progressed through several methods. The first method used the traditional branding iron in a wood fire, roping and dragging the calves to the fire, and throwing and tying the calves. A small improvement was made with the use of a propane fire holder for more uniform heating of the iron. We then converted to an electric branding iron—first powered by a gasoline generator and later powered by our solar electric system after it was extended to the corrals—thus the label *solar branded*. Finally, we converted to a freeze brand and have been using that method for the last 15 years (see figures 12.1 and 12.2).

Branding is the official proof of ownership in the state of Arizona and thus required. But many of the large beef-producing states don't allow it. Packing plants discount hot-branded cattle because the brand scar decreases the price they get for the hides. Freeze brands do not scar the hides because the branding affects only the color of the hair, which turns white.

Figure 12.1. The brass Soap Pot brand for the Quarter Circle U Ranch. This is the banding iron that we use to apply the freeze brands.

Figure 12.2. Family and friends help freeze-brand the calves at the ranch headquarters. Using the tilt squeeze chute, a.k.a. calf table, rather than roping, prevents stress and injury to the calves as well as the cowboys. The young man to the left of center is grandson Sean Doyle, and the young man branding the calf is Tanner Lannan, an extended family member. Photo by Judy Backus, April 2003.

At the QCU, freeze branding is performed with liquid nitrogen and a brass branding iron. After the calf is squeezed in the calf table, the table is turned so the calf is lying on its side on top of the table. The location where the brand is to go is shaved with an electric hair clipper. Removal of the hair allows good thermal contact between the iron and the hide. The brand area is then sprayed with 99% alcohol to again ensure good thermal contact with the hide. The reason that a low percentage of alcohol is not used is because it forms ice from the water content. Between brandings, the iron is kept in liquid nitrogen in an insulated container with a closed top, which reduces evaporation (see figure 12.3).

Figure 12.3. The handle of the brass branding iron, left, is sticking out of the Styrofoam container holding the liquid nitrogen at about minus 344 degrees Fahrenheit. The Dewar, right, is the large thermos bottle used for transferring from the 140-liter nitrogen Dewar that comes from the supplier. Photo by Jack Carlson, 2009.

After the alcohol is sprayed on, the iron is pressed against the hide at the shaved area for a total of about 40 seconds (see figure

12.4). Luckily the calves don't seem to feel this, as they lie perfectly still during those seconds. The cold (minus 344 degrees Fahrenheit) does not affect the hide but does damage the roots of the hair (the follicles), and thus the hair grows out with a white color.

Figure 12.4. A calf is secured on the calf table while Howard Horinek looks at his watch to time the freeze brand. Merlin Yeager, in the background, is holding the alcohol spray bottle. Photo by Jack Carlson, October 2007.

These white-colored freeze brands really stand out on dark-colored cows, and, of course, the brand size grows with the cow (see figure 12.5). Some people may think that the use of liquid nitrogen is exotic and expensive, but it is not. We rent a 140-liter Dewar full of liquid nitrogen, which lasts the entire branding season. The cost is less than $2 per calf.

For more about the calf processing facilities, see Chapter 10, "Headquarters Handling Facilities," in the section "Operation of the Processing Facilities."

Calf Processing in the Spring

Figure 12.5. This two-year-old Angus crossbred heifer wears the Soap Pot freeze brand of the Quarter Circle U Ranch on her left rib. She is identified by her yellow ear tag number 312. In the same ear, but not visible, she wears a small stainless-steel ear tag and an electronic radio frequency (RFID) tag. All newborn calves are tagged, recorded, and identified with their mother's tag number. Photo by Chuck Backus.

Electronic Identification (EID) Tags

Electronic Identification (EID) tags are attached to the left ear of the calf—between the ranch tag and the calf's head. The EID number is scanned with the EID reader, and along with the ranch tag number, they are saved on our laptop computer. These EID tags are numbered with a 15-digit number that is nationally recorded and identified with the QCU Ranch (see figures 12.6 and 12.7).

This EID tag stays with the calf right through slaughter, and thus if there is a problem with this calf or the meat from this calf, anywhere along the line, it can be traced back to our ranch. If there was a problem, we would get a call within 24 hours. This is part of the *traceability and credibility* program that the national beef industry is establishing.

On the practical side, the use of these tags makes processing cows and calves much quicker. The wave of a reader wand brings all the data up for a cow or a calf. The cost of a registered EID tag is about $2 per tag, and the cost of a numbered plastic ranch tag is about $1.25 per tag. We have not had a problem with the animals losing the EID tags.

Figure 12.6. The round button is the electronic identification tag (EID) with 15 numbers. The EID tags used on calves have numbers starting with 840 and are nationally registered to the Quarter Circle U Ranch for traceability and accountability. The inset shows a close-up of the EID tag. The plastic tag with number 649 is the ranch tag. The first number of the adult ranch tag is the last digit of the year she was born, i.e., all 600 numbered cows were born in 2016. The applicator can install both tags into the ear. Photo by Chuck Backus, 2017.

Calf Castration

Calf castration is always done on the calf table, which ensures that the calf is totally constrained—meaning the calf cannot move around to hurt himself or the worker. All bull calves are castrated, and all castrations are done with a sharp knife and an emasculator, which is the surest way to castrate. Iodine is always sprayed onto the area after the calf is castrated.

Heifer Calf DNA Samples

DNA samples are taken from all heifer calves using the Allflex tissue-sampling unit. Not all samples will be sent in to be analyzed but are available after the heifer calves have been visually evaluated over the summer and their sire and dam considered.

Figure 12.7. Chuck Backus, left, is preparing the ranch ear tags, and Howard Horinek, right, is preparing a spray bottle for applying the alcohol to the freeze brand area. Photo by Jack Carlson, April 2006.

Calf Vaccinations

All vaccinations given at branding are given in the neck as prescribed by the manufacturer and the Beef Quality Assurance (BQA) recommendations. Many vaccinations are administered subcutaneously, under the skin. All vaccinations are done while the calves are constrained on the calf table. This prevents accidental vaccinations of workers. Between vaccinations, the syringes are placed in an ice chest that is modified with tubes going into the side of the chest to hold the syringes.

The two vaccines given at branding are:

- Ultra Bac 8—an 8-way Clostridial Booster (black leg, etc.)
- Bovi Shield Gold 5L5—a MLV shot with a Lepto additive

Calf Dehorning

In the past, dehorning was required on some calves as the herd was being converted into a high percentage Angus herd. Dehorning was done at branding with an electrical dehorning tool over the horn buttons. It is extremely rare to find horn buttons now, so dehorning is normally not necessary.

Weaning

No one wants to buy bawling calves—even at the local sale barn. If we are able to let the cow wean her own calf, that is probably the best plan. If that isn't feasible, then we put the calves in a corral adjacent to the mothers, which is effective and is easier and more acceptable to both mother and calf.

Chapter 13

Cattle Operations: Artificial Insemination

An individual cow can give birth to 5 or 8 calves in her lifetime and influence their genetic characteristics. A herd bull could influence more calves than that per year, for a total of 30–100 calves during his lifetime in a herd. An artificially inseminated (AI) bull could influence several hundred calves in a herd. Thus, the primary way to improve a herd is with an emphasis on the live bull selection or AI semen selection.

There will always be a need for physical inspection of hooves, muscles, and structure to assure survival in a specific environment. However, there are better ways to select bulls than just their physical appearance.

The Case for Using Artificial Insemination

A technique for herd improvement that very few commercial ranchers (maybe 10%) use is AI. Most commercial ranchers consider AI a sophisticated technique that is available only to seed-stock operators because it requires very specialized handling equipment and talents. It does not. If you have a squeeze chute, that is the only equipment required. All the semen companies have local reps who can inseminate your cows. Those reps will order the semen you decide to use. You select which of the nationally known bulls you want to use, considering all the weighting factors you have used to select your live bulls. The reps or local trained persons will bring the bull semen to your ranch in liquid nitrogen containers and place the semen into the opening of the cervix of your cows and heifers. Many companies and universities have training programs to teach you and your ranch manager how to inseminate your cows. It is similar, and perhaps easier, than pregnancy testing.

University and industry researchers have found a way to cause many cows to come into heat at approximately the same time. They have several prescribed *protocols* that can be used to induce heat for several cows to be processed at one time, and thus up to several hundred cows can be AI-ed within a few hours. The development of these *fixed-time protocols* is what makes it practical for commercial ranchers to use AI on the ranch.

The downside of using AI is that it requires more processing of cows and thus ranch labor, as opposed to just turning the bulls out. The total process requires the cows to go through the squeeze chute 3 times, which may require the cows to be fed for up to 10 days during their time in the corrals. A rancher should ideally put the cows in a holding pasture for those 10 days, which would reduce the feeding costs.

The economics of the use of AI indicate that the cost of a calf on the ground is about the same for either AI or a natural bull. The biggest difference is the ranch labor required for AI and the impact it may have on annual ranch processes. However, if a rancher AIs many cows, they could reduce the number of bulls required in the herd by up to a half. AI allowed us to use higher-quality bulls than we could afford to buy, and we didn't have to maintain the bulls and keep them alive on the ranch.

The percentage of cows becoming pregnant from AI, administered in a ranch environment, is typically between 40% and 75%. My personal experience is that it varies between 45% and 60%, depending on the range conditions. Cows can be turned out with the bulls on day 1 of the breeding season, when we know that about 55% of my cows and heifers were bred to a very outstanding bull. Thus, AI allows ranchers to progress faster in the direction of their goals and inject superior genetics into the herd.

Artificial Insemination: Fixed-Time AI

Fixed-Time AI is an artificial insemination (AI) protocol intended to interrupt the cow's normal cycle in such a way that she can be induced to come into heat within a certain time interval. This procedure avoids the need for heat detection and makes AI practical for a commercial ranch. The protocol allows many cows to be bred by AI in a short time.

In the spring of 2017 at the QCU Ranch, on day 10 of the AI protocol, 200 cows and heifers were bred in a 4-hour window. During this 10-day Fixed-Time AI protocol, the cows need only be processed through the squeeze chute three separate times with shots given each time.

The first time through the squeeze chute (day 1), she is inserted with a vaginal appliance called a controlled internal drug release device (CIDR) (see figures 13.1 and 13.2). The CIDR is a plastic frame coated with progesterone, which is activated by a warm and moist environment. After the CIDR is inserted, it will stay in place for seven days before her second time through the squeeze chute. A vaccination is also given on day one to help start the process.

Since the cows must wait for seven days before the next step in the process, the cows and calves need to be corral-fed or turned out to a convenient pasture. Over the years, as the number of cows being AI-ed increased, two separate canyons near the corrals were fenced to hold the cows for these seven days. These two canyon pastures are heavily used once or twice a year for this purpose and rested for the remainder of the year.

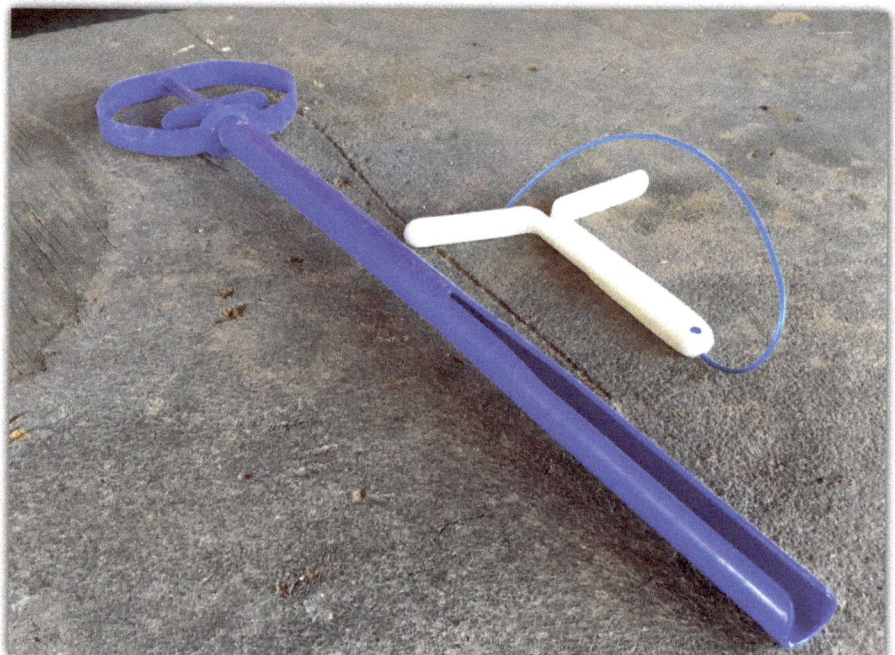

Figure 13.1. The controlled internal drug release (CIDR) is about 6 inches long. It has a plastic frame that is coated with progesterone that is slowly released when placed in a warm and moist environment. It is inserted into the vagina of the cow to help reset her normal heat cycle. A special plastic applicator is used for insertion. The top two wings are pulled down to the central structure for insertion into the applicator. After placing the loaded applicator in the cow, a plunger on the applicator pushes the CIDR out of the applicator, and the two wings come back out to help retain the CIDR in place for 7 days. Photo by Jack Carlson, 2009.

For the cow's second time through the squeeze chute (day 7), the CIDRs are removed, and a different type of shot is given that tells her body to proceed with going into heat. The cows are now kept in the headquarters corrals. While the cows are in the corrals for those three days, the calves of those AI mother cows are branded and processed.

On day 10, which is about three days later (66 hours, plus or minus 2 hours), the cows are artificially inseminated by inserting the bull semen, which is followed by a third shot. This completes the Fixed-Time AI protocol, and the AI-ed cows are turned out into the pasture with the live bulls.

Cattle Operations: Artificial Insemination

Figure 13.2. This photo shows the CIDR loaded, with the wings collapsed in the applicator. Photo by Jack Carlson, 2009.

Chuck is preparing the CIDR. Photo by Jack Carlson, April 2006.

Chuck is preparing the CIDR. The gate located at the back of the squeeze chute is behind him—white posts. This gate allows the operator to enter and stand in the stall behind the cow and perform the AI-ing. Photo by Jack Carlson, April 2006.

The cow is secured in the squeeze chute, and Chuck Backus, right, is inserting the CIDR with the special plastic applicator while ranch manager Howard Horinek, left, assists by placing a bar behind the cow's legs to keep her from accidentally kicking Chuck. The CIDR will stay in place for seven days. Rubber gloves are worn when handling CIDRs for sterility and to ensure no human contact with a hormone-containing substance. Photo by Jack Carlson, April 2006.

The Fixed-Time AI protocol has been used at the QCU Ranch for more than 10 years and typically results in about 55% of the cows becoming pregnant from the bull semen used in the AI. Thus, more than half of the cows will become pregnant on the first day of the breeding season, which greatly enhances the predictability of the calving season. Universities, with more controlled conditions, report up to a 75% success rate.

Scheduling the AI Protocol

Thirty days after the cows give birth, they are ready for AI processing. We AI the cows over a three-month period in the late winter and spring, and their calves are born at the beginning of next winter. We have three or four distinct groups of cows go through the AI processing to ensure that most of the cows and heifers get bred by AI.

Before the process starts, it is recommended that the cows have calved at least 30 days prior to inserting the CIDR. Since cows have a nine-month gestation period, there is a chance to move cows up in their calving intervals by a month or so each year. Early calving is preferable because calves will have more time to graze in the winter and thus gain more weight. Heavier calves are desirable. Early calving also indicates that the mother is more fertile. We plan to have the first group of AI-ed cows be as large as possible.

In the spring of 2020, more than half of our cows were AI-ed in the first group. It is recommended that the heifers go through a longer protocol than the 10-day cow protocol, but they can be processed on a time schedule so that they are AI-ed on the same day as the first group of cows. This allowed us to schedule 200 cows and heifers for artificial insemination in a 4-hour window.

As we rounded up more cattle, we grouped the cattle that had 30-day-old calves, and when we had a group of about 20 cows, we scheduled them for the 10-day protocol. These smaller groups were

AI-ed by our ranch people. AI-ing cows is physically demanding, so we kept the groups small that were processed by our ranch people.

The ability to process cows for AI at the ranch using standard ranch facilities is very straightforward and can be done mostly by ranch personnel. The exception is the experienced person that deposits the bull semen inside the cow's cervix. The person that inserts the semen requires special training. Most local semen-distributing companies have a trained AI technician who will do the insertion at the ranch for about $5 to $6 per cow.

AI Technicians

The AI technician sometimes comes out to the ranch with a *breeding trailer* that can be backed up to a normal squeeze chute. It has two compartments so that the ranch people can be loading one side while the technician is AI-ing a cow on the other side. This obviously speeds up the process. However, we AI-ed 200 cows in four hours using our normal squeeze chute with a professional technician. Our ranch managers at the time (Dean Harris and his wife Kris) had both attended the Graham School in Kansas and could breed up to about 50 cows in one day. It takes a strong arm to breed a larger number of cows. Currently, we use a professional technician for the first large group and our ranch managers for the other three groups.

First Evaluation of AI and New Goal Set

The use of AI was first tried on 50 cows at the QCU in the spring of 2006 to see if superior bull calves could be raised on the ranch by their native mothers. The conditions on the ranch are so harsh that bringing in outside bulls or cows often resulted in them dying. They get sore feet from the rocks and do not know what to eat. By using AI bull semen, we can affordably select the best bull in the country to meet our requirements without having to buy him or worry about keeping him alive.

This successful AI experiment resulted in a decision in 2007 to raise our own replacement heifers from AI and change the goal of the ranch to be: *Raise high carcass quality calves and benefit from them by retaining ownership of those calves through the feedlot, selling on the grid, and receiving other premiums at the packing plant.*

To replace the entire herd with improved mother cows is an exceptionally long process. We cannot just buy better cows. They would die from the harsh conditions if they were turned out. However, using AI allowed us to speed the process up. We knew that it would take a minimum of 10 years to significantly change the herd. The people at the Certified Angus Beef (CAB) organization were helpful in providing information and advice. Our new goal is in line with their requirements for certifying carcasses as CAB quality.

Retaining Yearlings for Future Herd Cows

Although I wanted to send the higher-quality marbling calves to the feedlot as yearlings, I needed to retain some of those animals to improve the herd. Using bull AI semen in my existing cows would not achieve my goal very quickly because the existing cow genetics obviously controls half of the genetics of the resulting calves. Thus, I retained the better female calves born from the AI-ed cows and added them to the herd. As those calves grew to become herd cows, they were processed with AI semen, which produced better calves, and some of those calves were added to the herd. Adding new calves to the herd is a slow process that never ends but produces a better herd each generation.

Of course, we wanted our cows to have additional characteristics other than just the high marbling, and thus we had to compromise and lose some marbling characteristics to gain other characteristics such as height and stamina, which are important for our cattle to live on the rugged terrain of our ranch.

Live Bull Selection

In April, during the time when the cows and heifers are being AI-ed, the annual bull sales are being held. The QCU buys live bulls from the Bar T Bar Ranch near Winslow, Arizona, which primarily sells registered Balancer bulls but also some registered Angus bulls; we also buy from the Midland Bull Test Center, Montana, where we bought both Angus and Balancer bulls; and the Leachman Cattle Co. of Colorado, where we bought Angus bulls.

The decision on what bulls to buy is almost entirely based on numbers that are provided by the seller—intermuscular fat (IMF), residual feed intake efficiency (RFI), and expected progeny differences (EPD). We buy the bulls on their numbers—not on their looks. We limit our bull purchases to the three bull sales listed above because of the large number of bulls they test and the reputation of the companies. Most of the live bulls we buy come from the Bar T Bar Ranch, and we partner with them in the purchase of bulls and bull semen. The bulls are bought when they are yearlings to better ensure that they can adjust to our rough country.

All of the existing bulls in the QCU bull battery are either Angus or Balancer bulls that are more than 50% Angus. All bulls bought must have been tested for a minimum value of marbling and residual feed intake efficiency. Marbling is sometimes called intermuscular fat (IMF), and it is a measure of the amount of fat that appears in the body of the animal. Marbling is a major contributor to a steak tasting good and juicy. Thus, we want the marbling characteristic to dominate and to be on the higher side of normal.

Residual feed intake (RFI) is the amount of feed required for a steer or heifer in a feedlot to put on a pound of weight. This number is desired to be low, thus requiring less feed to fatten an animal while in the feedlot. The sale books give the residual feed intake efficiency (RFI) for each bull as well as the average for the herd. The DNA-enhanced expected progeny differences (EPD) must also be available for the bulls. Chapter 15 has more information on DNA sampling, testing, and analysis.

Our major selection criteria for live bulls are based on the residual feed intake, marbling, the index numbers that are a combination of the economically important EPDs, and the bull being black. Bulls with genetics that predict low milk numbers and high yearling weights are preferred.

We also look at the bulls' physical characteristics to see if they have decent feet and structure. These physical characteristics are important factors for the cattle to survive our harsh desert environment and for the premiums paid for our calves.

AI Bull Semen Selection

The evaluation and selection of AI bull semen to be used at the QCU Ranch is a process that continues over the entire year. Data is collected from reading many articles and magazines on bulls and genetics, attending several meetings and conferences, evaluating several semen catalogs, and communicating with several cattlemen in the industry.

The main advantage of using AI on the ranch is that we can select from literally thousands of bulls and thus have a better chance to select the bull with the characteristics that we want. One cannot afford to buy bulls that are as good as the bull semen for bulls available for AI. Suppliers can obtain more than 1,000 AI samples from one ejaculation from a bull. Thus, the cost is very low per sample, maybe between $5 and $25. Depending on the bull, they can get several ejaculations per day. They use the very best bulls in the country, so one cannot buy better bulls.

The other advantage of AI is that we don't have to keep that bull alive in the rough country of the QCU Ranch. For very popular bulls, their semen is available long after they are dead. Some AI bulls have thousands of calves with recorded characteristics, and thus their EPDs, adjusted with the data from their calves, are extremely accurate.

We buy *straws* of semen and use one straw per AI. These straws of semen are sealed plastic straws about 1/8 inch in diameter and 5 inches long, which contain more than enough semen to fertilize one cow. Assuming only one calf is born for every two cows that are AI-ed, we can get a possible *super calf* from a bull that we could never afford to buy for perhaps a total processing cost of $50 to $70.

We have also acquired a commercial liquid nitrogen Dewar, which allows us to keep straws from different bulls over the entire year. We sometimes buy a percentage interest in a new bull, where someone else must keep the bull alive and send us perhaps 100 straws each year.

In the early days of using AI at the QCU, a main criterion for selection was for bulls to have high calving ease and low birth weight. That was because the calves were born in remote canyons and thus unassisted. That is less important now since the mother cows have that characteristic. It was also important to initially use Angus bulls with very high marbling to build up the quality of the calf carcasses. Now the selection criteria for semen are very similar to that for live bulls, just more demanding.

Another advantage of AI is that we can match up an individual bull to an individual cow. This allows potential deficiencies in a cow to be compensated for by a bull that has traits that are strong in the cow's deficiency areas.

The use of AI can also ensure that we minimize the chance for inbreeding. No AI bull semen will be used on any cow that is related to that bull. AI use also allows us to change bulls every year, if desired, and use a greater variety of bulls over the years.

In the spring of 2017, semen from six different bulls was used for AI. Different bulls were used for specific reasons on different cows.

For 2021, the data from the feedlot and packing plant showed the quality of the carcasses was high. The steers graded more than 50% Prime and 94% either Prime or CAB quality.

Now, with the high quality of our animals, the criteria for semen selection is weighted higher for feed conversion efficiency (FCE). High FCE should not only require less feed in the feedlot but also reduce the amount of feed required on the ranch to maintain the mother cows and allow them to be in better condition to breed back after calving.

Documentation

It is very important that the processes discussed above are recorded and documented. Documentation does not have to be a demanding or time-consuming effort. We use a simple, spiral-bound notebook like the kids use for school. Any time we are working cows in the corrals, we record the following: the date, the process being done, any vaccines used, the number of cows processed, and any data generated, i.e., the cow tag number, her pregnancy level, or other information that may be useful later. A single notebook will last for two or three years. It is surprising the number of times we need to go back and check what we had done with a certain cow or calf. We all think that we can remember all that information in our heads, but it is very human not to remember all the details. Our ranch managers usually do the recording and later enter the data into the herd computer program. Sometimes they enter it directly into the computer at the squeeze chute.

Today, almost all ranchers use email. It is just as easy to use a program like Microsoft Excel to make a spreadsheet of the information for all of your cows. If a rancher needs help getting started, another person can initially set up a herd data program that the rancher can update and use. The reason most ranchers are reluctant to try new things is because "We have always done it this way, and it works for us."

Cattle Operations: Artificial Insemination

Moving the cows and calves down the Tule Trail from Tule Divide for processing at the ranch headquarters located just beyond the left edge of the photo. Photo by Jack Carlson, April 2005.

There have been tremendous changes and improvements in the United States cattle industry over the last 10 years, and Western ranchers need to use that latest technology to compete with the Midwestern farmers. Unless Western ranchers use more modern techniques to improve their quality, the price paid for Western calves may decrease.

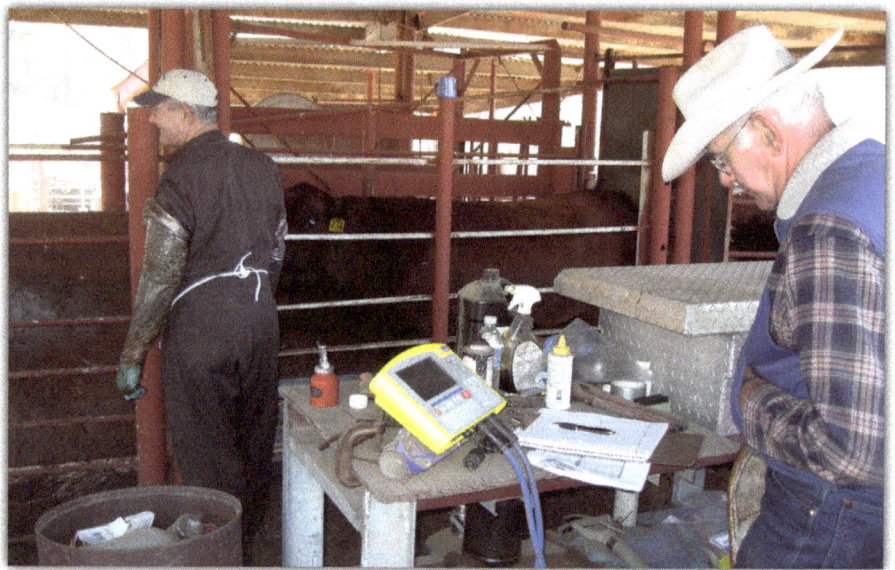

Chuck Backus, right, is looking at the recording notebook and the results of the weighing. The yellow device is the display for the weight scale. Cow 434, in the queue line, is on the scale. Veterinarian Larry Lunt, left, wearing the shoulder-length gloves, has been performing the pregnancy evaluations.

Chapter 14

CATTLE IMPROVEMENTS: GENERAL CONSIDERATIONS

The first thought for how to increase the income from our operation was just to increase the number of cows and calves we produce. However, previous generations considered that idea and had already established the number of cattle that a given ranch could support. Since a lot of Western state ranches primarily operate on publicly owned lands, the respective government entities determine the actual carrying capacity of cattle permitted on a particular parcel of government land. Therefore, a rancher's usual options are to improve the feed at the ranch or increase the value of individual calves—or sometimes both. Both of these options are possible at the same time, but both take a long time to implement. One option involves range management, and the other involves genetics.

To make these types of improvements, it is probably best to set up and evolve into a yearly schedule for the entire ranch. This entails rotating pastures in a prescribed way and setting certain periods of time for breeding cows, calving, and weaning, and for turning bulls in with cows. Cows seem to adjust to yearly schedules.

Need to Learn About Tools

To use more effective ways of improving our herds, we first must understand the tools available and how to apply them. Many commercial ranchers are not even familiar with the terminologies used, much less how to apply them. Therefore, we need to learn more about basic genetics and how we can use these powerful selection tools that are available. In addition to learning about the selection of bulls to add to our herds, we need to consider using artificial insemination as a practical tool for commercial ranchers to use. This allows the use of semen from the best bulls in the country—which most of us can't afford to buy for our personal herds.

The research and understanding of the area of bovine genetics is developing so rapidly that it is hard for seedstock producers to keep current and much more difficult for commercial ranchers. The land-grant universities in the United States were established and funded to provide this leadership. But even ranchers who have graduated from these schools are hard-pressed to keep up. The cattle breed associations and private companies are also doing research and trying to develop tools to incorporate these new results into practical tools for application by commercial ranchers. Every year the industry has new and valuable research results to incorporate into producing a better opportunity for ranchers to improve their herds.

Beef Grading

Progressive commercial ranchers have been using these genetic tools in the last few years to increase the quality of beef—making it a better eating experience. Traditionally, the percentage of beef coming out of packing plants has been graded 50% Choice or better and 50% Select. In 2018, US beef averaged 80% Choice and better. The beef grading Prime went from the traditional value of 3% Prime to 9% Prime in

2018. The percentage of Select in 2018 went down to 20% from the traditional level of 50%. There is now talk that the USDA may even drop Select as a beef grade within the next five years.

Records and Measurements

A producer cannot have too many records of individual cows and bulls. Cattle, like humans, are individuals with individual characteristics, and some of the past generations are not necessarily predictive for individual animals. For example, there seems to be a common belief that there is a relationship between birth weight (BW), yearly weight (YW), and mature weight (MW). However, there are individual cattle that have low birth weights (highly heritable) and high yearling weights. Another example is that there is a general expectation that high-milking cows have higher-weaning-weight calves—which may be generally true. But there are cows that have moderate to low milk production but have high-weaning-weight calves.

Peter Drucker, a prominent industrial engineering professor, always emphasized, "If you can't measure it, you can't improve it." A rancher needs to start measuring the characteristics of their cattle, find out what the buyers will pay more for, and then improve those characteristics. For example, most of us know that heavier calves bring a higher total price, but how many ranchers have a weight scale for individual calves and know which cows produce the heavier calves and which bulls those better calves came from?

Development of EPDs

Before the turn of the last century, numerical measures to quantify and compare various characteristics of cattle started to emerge from purebred producers and beef associations. They developed expected progeny differences (EPDs) that predict the characteristics of the progeny (their calves) for individual bulls. A commercial producer can compare these numbers for different bulls (for example, what their calves would weigh at birth) to make better-informed selections. Generating EPDs requires a tremendous amount of data collection and analysis. A laborious effort of record-keeping by purebred producers is required as well as the breed associations and the availability of large computers to analyze that data.

Setting, Measuring, and Commitment to Goals

After setting a goal, or even before setting a goal, ranchers should establish the current characteristics of the existing herd, especially the traits targeted for improvement. That way, you can annually monitor the progress toward the goal set. Also, ranchers need to understand more about cattle traits and which ones are typically available for consideration in bull and replacement heifer selection.

Before we use the tools, we must figure out what we want to accomplish by using them. We must determine what characteristics we want our cows to have and then set both ranch and herd goals. An old adage says, "If you don't know where you want to go, then any road will get you there." Thus, each of us must decide by our goal or target the ultimate characteristics that we want our cattle to have. In determining that goal, we must consider what the physical limitations (terrain and climate) our ranches have, determine what it would cost to move toward the goal, and determine if your family has the desire and the personal commitment to meet the goal. In addition, we need to consider where and how we market our calves or where we could better market our calves if we meet the goal. For most commercial ranchers, the goal set will be based on the economic impact the attainment of the goal will have on their overall ranch operation. All these considerations must be made on an individual-ranch basis.

To determine the progress toward a goal, a rancher may have to obtain, by measurement, additional data on their cows and especially on their calves. Ranchers can, for instance, arrange with a feedlot to obtain the individual calf-carcass data from the slaughter of their calves—even if they are sold to a feedlot buyer. Another factor to consider is that consumers are becoming more interested in where and how their beef is raised. Therefore, ranchers may need to document what months their calves were born, the type of feeds they were given, and the injections they received. The measurements that are needed will depend upon the breeding goals.

A significant amount of time will be required to change the characteristics of a herd. A breeding goal is used to identify the traits targeted to be improved, along with the amount of emphasis on each trait. A breeding program is aimed at the next generation of the herd. The attainment of the goal may take several cow generations, so a rancher should plan for continuous improvement rather than expect to quickly meet the goal. In addition to bull and replacement heifer selection, a rancher might consider a culling program that could eliminate the existing cows in the herd that do not have good traits that are targeted for improvement. For example, culling cows with poorly performing calves.

The biggest time constraint is the gestation period of the cows and the two years before the calves can produce the next generation of improved cattle. Cows need to be about two years old when they have their first calf. A mother cow's life in the herd is about 10 years, although cows can live for more than 12 years or more. It took 10 years for the herd to improve enough to meet my goal.

My goal was to *produce calves that would grade Prime or Certified Angus Beef quality cattle, both steers and heifers, coming out of the packing plants.* I was 90% successful in finally getting there before I retired. Those first 10 years after I decided on the goal saw the biggest improvement in the quality of the calves. There is no limit to adjusting and implementing the goal. The process can continue until all cattle coming out of the packing plant are graded Prime.

Bull Genetics

The cattle industry has changed very rapidly during the last 20+ years, primarily because of the advancements in the understanding of cattle genetics and the development of expected progeny differences (EPDs). Appendices B and C have additional information on EPDs.

For the last few hundred years, the *best breed* and the *best crossbreed* of cattle have been factually and emotionally debated. There are certainly typical characteristics of breeds, but there is more variation within a breed than between breeds. Cows are individuals.

Today bull sales are dominated by measurements made on individual bulls, including DNA results and predictions for the characteristics of their calves (EPDs and DNA-enhanced EPDs). Even with today's knowledge of genetics and measured characteristics, most commercial cattle people select their bulls mostly on their physical appearances. One hears buyers say, "I really like the looks of that bull," or "I want all of my calves to look like that bull." This follows what people have used for several hundred years with the assumption that "like begets like." If one only has appearance as a means to judge, that is not a bad assumption. But it was used because people did not have today's understanding of the field of genetics and the measurements that we have available today.

Genome Development and Enhanced EPDs

The bovine (cattle) genome was developed in 2009, just a few years after the human genome was developed. The *genome* identifies the total possible helical structures (the basic building blocks of life—the DNA) that exist in each species. This is sometimes compared with written languages that first require all the letters used in a language to be identified before combining them together into meaningful words or sentences. After all structures are identified, researchers can try to identify which group of structures, such as which genes determine a certain characteristic of the species, i.e., marbling in cows. Thus, knowing which DNA structures exist in each bull can better predict what characteristics his progeny might inherit. This is an extremely powerful tool, and it is constantly getting more accurate and easier to use for a commercial beef producer. Ranchers can also do DNA testing of potential replacement heifers for better-informed selection of replacement heifers that better meet the goal for the future herd.

Cattle beef associations are now combining the traditional EPDs that were originally used, based on measured characteristics of their calves, with the DNA data for an individual bull to make even better predictions of the characteristics their calves may have. These combined data of both measured and DNA predictions are called genetically enhanced EPDs (GE-EPDs). Breed associations usually generate the GE-EPDs.

The bull and replacement heifer selection tools in the beef industry are getting better and more accurate all the time. A commercial producer may soon have tools available that one can almost use to design the characteristics of the herd they want and then select bulls and replacement heifers to meet that design. Also, the now common practice of artificial insemination (AI) allows one identified super bull to produce thousands of either super male and/or super female calves.

Cattle Traits and Heritability

Cattle traits are classified both as qualitative traits that are observable and/or measurable and as quantitative traits (multigenetic) that require DNA analysis.

Qualitative traits are traits that were usually used in the past that are observable or measurable, such as hair color, horned or not, bone, muscle, and foot structure; adult size and appearance—both for the sire and dam; as well as measurable data such as birth weight, yearling weight, etc. Qualitative traits are usually described as defining the *phenotype* of the bull's traits.

The quantitative, or genotype traits, describe the genetic make-up of the bull, which requires DNA testing. Both types of traits are important to provide information for comparing different bulls. Both types of traits describe the bull, but the probability of him passing those traits on to the progeny (his calves) is determined by what is described by the term *heritability* (denoted as h-squared). Heritability is the probability (0–1) of inheritance, and it varies for each trait in both the phenotype and the genotype of traits. Plus, it is different for various traits. Each calf receives half of its genes from the sire and half from its dam. Thousands of different genes exist in each animal, but since only one member of each gene pair comes from each parent, the calf's inherited genes are determined purely by chance. Thus, the calves from the same parents vary in inherited characteristics but are similar. In our personal experience, we know that siblings are often quite different from each other and sometimes even different from their parents.

The detailed definition of expected progeny differences (EPDs) and their use was explained at the workshop I organized in February 2022 by Dr. Matthew L. Spangler—a professor at the University of Nebraska-Lincoln. A copy of this presentation appears in Appendices B and C.

Feedlots Want Healthy Calves that Bring Premiums

What happens after our calves leave the ranch affects the price we receive. Commercial ranchers need to learn more about the rest of the cattle industry to better prepare their calves to receive the best price and premium for their animals.

Essentially all calves raised by ranchers, except for replacement heifers, go to a feedlot. Thus, we should consider feedlots as the customers for our product. A cattle buyer acts as the intermediate owner—buying calves from ranchers, often at the local sales barn. The price that a buyer can pay for your calf depends on the price they think they can get for your calves from a feedlot. Cattle buyers get to know the types of calves coming from various local ranchers and have some idea which feedlot will buy these types of calves and the price they can probably sell them for. The feedlot is what determines the base price the buyer can pay for your calves. Thus, we should prepare our calves for what a feedlot wants.

Calves need to be vaccinated against major cattle diseases that a calf may bring to a feedlot that could spread to other calves in the feedlot. Also, since calves in that feedlot come from other states, they need to be vaccinated against those diseases. The feedlot can advise on the needed vaccinations.

Some calves may get sick or die in the feedlot. If they die, then that is a direct loss to the feedlot if the feedlot owns them. If they get sick, the feedlot must remove the calf from a given pen and put them into a *sick pen*. In the sick pen, they have the expense of giving them shots, typically antibiotics, plus the calves are set back in their weight gain. A feedlot is lucky if it can break-even on a calf that goes into a sick pen and certainly loses money if the calf must enter the sick pen twice.

Calves need to be completely weaned and backgrounded before going to the feedlot. When calves are taken directly from their mothers, they have an emotional transition as well as an adjustment from milk to some type of new feed. Calves typically lose weight while they make this transition and usually require several vaccinations or booster shots during this time. Backgrounding is best done on the ranch, which is less costly than having the feedlot do it. If the calves are not backgrounded, cattle buyers must take this into account and lower the price they pay for those calves accordingly. If the buyer does not background the calves and sells directly to the feedlot, the buyer will receive a discounted price from the feedlot for those non-backgrounded calves.

Feedlots prefer calves from ranchers that have had their calves in the feedlot before, and they will have kept records on past performance on calves from that ranch.

Buyers prefer types of calves that they think feedlots and packing plants, their customers, will pay premiums on.

Feedlots prefer calves that have been certified by an official third party because there is less uncertainty in the quality of the calves.

Feedlots prefer and almost require calves that have been properly backgrounded.

Feedlots make most of their profits from the selling of feed for calves in their feedlot. The owners of the calves in a feedlot include the feedlot, an investor that buys calves with the hope that they can make a profit on those purchased calves, a rancher who wants to make an additional profit on the calves they have raised, or a company that is integrated by owning ranches, feedlots, and sometimes, even the packing plant that processes and sells beef. Feedlots vary, depending on their business plan, from owning 100 percent of the calves in their feedlot to owning none of the calves they feed. Since the expenses of a feedlot are determined mostly by the cost of feed, most feedlots are located near feed sources, for example, the US Midwest states. Feedlots vary in size from a few hundred head capacity to a few hundred thousand head capacity. Fattened cattle go from feedlots directly into packing plants.

Packing Plants

Large packing plants take in fattened cattle from feedlots and process them into *boxed beef cuts*. The boxes will have only one type of cut of beef, such as ribs or brisket, and they will all have the same grade but from different animals. These boxes will be wholesaled out to grocery stores, restaurants, or distribution companies. Packing plants typically process 1,000 to 5,000 head per day with crews that run two or three shifts per day. Cull cows and bulls are usually processed in many smaller packing plants that are more widely located because their sources of cattle are more widely scattered.

Typically, fattened cattle, weighing 1,100 to 1,500 pounds, are delivered to a packing plant by truck, individually weighed, and priced at the *fattened cattle commodity price* on that day.

If the feedlot has requested that the cattle are to be *sold on the grid*, the cattle are individually weighed. The cattle are slaughtered, cleaned into carcass halves, and hung in a cooling room overnight. The next day, the halves are moved into a grading room where the side is cut open between the 12th and 13th ribs, and the ribeye is measured for area, marbling, texture, and amount of measured external fat layer. Also, a USDA representative inspects for health purposes. The quality of the entire carcass (subsets of Prime, Choice, Select, etc.) is determined by the measurement of marbling in that one ribeye area.

The quality grading was traditionally done by an expert human inspector. After many years of testing, they are now mostly using cameras with computer analysis—thus the testing is much more detailed, reliable, and consistent. (Cameras and computers never come to work with hangovers.) The amount of marbling in a ribeye varies gradually and varies from none to excessive. Thus, the grading identifies the carcass as belonging to one of the three general grades, such as Prime, Choice, and Select, and then it divides those grades into many subcategories. If requested and paid for by the carcass owner, the detailed quality grade can be reported for the many subcategories within a general grade for each individual carcass. The number of subcategories in each of the general grades is Select (10 sub-grades), Choice (30 sub-grades), and Prime (20 sub-grades).

Since there is such a large range of sub-gradings in Choice (30 sub-grades) and a lot of carcasses fall within that grade, Choice is also graded and grouped into three more general categories: High Choice, Medium Choice, and Low Choice. Each of these three general categories has 10 sub-grades. There are many more carcasses that grade in the Low Choice than carcasses in the two upper categories combined, so most grocery stores label any cuts in Low Choice as just Choice and all carcasses in the upper two categories as Premium Choice.

If the carcass owner requests being paid *on the grid*, then, in general, the price paid, based on the live animal weight, is greater for a higher grade. For example, we often read in a market price report about the Choice to Select Spread, which is the difference between the price paid for Choice carcasses and Select carcasses on a per-pound basis of the live weight of the animal coming into the packing plant. There is a major jump in the price for carcasses that grade Prime. Of course, the reason that the packing plant pays a premium for these higher grades is because they can sell them for much higher prices. Many packing plants will also give an even higher price for Premium Choice grade carcasses. Since Certified Angus Beef (CAB) requires all their certified beef cuts to be Premium Choice or better, some packing plants call this category CAB Grade.

We receive the quality grading results of the 18-month-old calves when they are slaughtered and request and pay for the carcass data results. Each carcass is graded, and the grade determines the price we receive for the carcass from the packing plant. I thus retained ownership through the packing plant because that was the first place in the process that I affected the sale price. The premiums paid for higher-quality carcasses can provide ranchers with an additional consideration, perhaps up to $300 per head.

Summary

Ranchers need to understand more than their own ranch operations. To be profitable, ranchers must understand the complete beef industry as shown in the flow chart in the introduction of this book. Familiarity with the seedstock breeders will get you started, and knowing what the consumer wants will help you set the herd goals. On-the-ground research by ranchers is an easy way to increase your knowledge. Producers need to visit feedlots, packing plants, and retail outlets and discuss their needs and also their consumers' needs.

Howard Horinek on the Coffee Flat Trail coming into the Reeds Water area. Howard is riding Little Buddy and leading packhorse Clem, who is loaded with tools for maintenance work on the springs in the Fraser and Tule Pastures. Cow dogs are Ruff, Spike, Gus, and Tuff. The dogs are a big help in moving the cows and holding them up if we need to get around the cows on the trail. They are very smart and energetic. Buzzards Roost is on the right horizon. Photo by Jack Carlson, October 2007.

Chapter 15

HERD IMPROVEMENT: BULL SELECTION AND OTHER METHODS

Some of the most economically important selection traits for bulls are not very heritable and/or are difficult to measure. Those traits include feed conversion efficiency (FCE), reproduction and conception, days to calving, stayability (STY), and the 30-month pregnancy rate. For example, the feed conversion efficiency trait is what determines how many pounds of feed it takes to put on a pound of weight in the feedlot and how much natural feed a cow needs to consume to maintain her body condition and breed on an annual basis. Not only is FCE difficult to measure, but it also varies drastically, perhaps by a factor of two to four, from one animal to another in the same herd. For a bovine, the genes or markers within the genome that control this trait have not been identified. However, that trait is only mildly heritable.

Selection Indexes

Commercial ranches vary a great deal on how they operate and what product they produce, so they may find it challenging to select the proper bulls that have the greatest fiscal impact on their ranch.

Genetic experts have attempted to help simplify the complications of selecting bulls for the commercial rancher. They have generated indexes with a combination of traits and weighted them by their estimated economic impact. These parameters are called selection indexes, or sometimes called dollar indexes, such as $B, $W, $Cow, etc. Trait selection based on these indexes may lead to a faster budgetary impact on a ranch. Ranchers inexperienced in genetic bull selection may initially want to start by using these indexes. Then later, a rancher may want to select especially for their chosen traits to better meet their herd goals, as well as consider the dollar indices.

Prioritize Several Traits

Besides deciding which traits are to be used to select a bull to meet a rancher's goals, they need to evaluate the technique or process for the selection for these traits. For example, it is not recommended to select only one trait and ignore all the other traits of a bull. That could lead to a disastrous result. The producer needs to choose and prioritize perhaps two or three traits that are to be the focus for meeting the herd goals and then perhaps set acceptable and allowable values for the second and third tier traits. The producer may also have to decide to buy or use bull semen that has minimum or maximum values on certain operational traits, for example, calving ease.

Physical Environment

The very first step in bull selection needs is to determine what limits are imposed by the physical environment at the ranch, such as terrain, climate, and feeds available. Cows are very adaptable, but it is difficult to take a cow or bull that has been raised in one environment and have them adjust quickly to a drastically different environment.

Numbers Based Decisions

Selecting good bulls to meet certain goals requires processing the numbers. It does not require math but just requires comparing and making decisions based on numbers. Not all ranchers are comfortable working with numbers. All of the data recorded on the characteristics of bulls are in numbers. In my case, I had Ag roots, but I was educated to be and spent a professional career as an engineer. Thus, I am very comfortable working with numbers.

Setting First Goal and Secondary Traits

In setting a goal, first you should select the herd goal for characteristics you want your ideal herd to have—perhaps two to four traits. They should be reasonable to achieve within a few herd generations. Then, you need to prioritize those characteristics and decide which set of measured herd characteristics you will use to measure your progress toward the goal.

After choosing your primary goal for your herd characteristics, you need to select additional, secondary traits you want to include in the characteristics of bulls you select—for example, birth weights. In my case, in 45 years of ranching, I have never assisted in a calf birth. My calves are born in remote canyons; therefore, I want my cows to successfully calve unassisted. This means that I need to choose my bulls that have EPDs with high calving ease (CE) and low birthing weights (BW). It turns out that CE is a better EPD to emphasize than BW. The calving ease is determined more by the pelvic structure of the cow than the actual birth weight. Also, actual birth weights may vary with different environmental conditions, such as rainfall.

In bull selection, the EPDs for secondary characteristics may just require setting minimums or maximums on allowed traits or just stipulating that those traits must be above or below the averages for that breed. For example, I want the calving ease to be above average for that breed. Some reported traits for bulls may not even be considered—especially when you start on your selection journey. After you begin to approach your primary herd goal, you may want to start selecting additional traits.

EPD Comparison Within the Breed

Also, in selecting certain bull characteristics, the producer needs to not only look at the EPD number of a bull but also how that EPD compares with all the bulls within the breed for that characteristic. Thus, many bull descriptions illustrate how that EPD number compares with all the bulls in that breed. For example, a bull description in a bull sale catalog might report an EPD of an Angus bull for marbling as 0.80 (25%) or as 0.49 (60%). This means that the first bull has a marbling EPD of 0.80, which ranks him in the top 25% of the Angus breed; the second bull has a marbling EPD of 0.49, which ranks him in the top 60% of the Angus breed. The various breed associations usually produce a table each year, showing the EPD numbers and how they vary for each EPD as far as which percentage that number ranks in the percentile within the breed. They do change, and thus the producer needs to use the current year's table.

DNA Testing for Replacement Heifer Selection

Herd improvement can perhaps most effectively be achieved by bull selection. However, a rancher can also make significant improvements by carefully selecting replacement heifers—especially by not keeping those heifers that may appear physically desirable but not genetically desirable. There are now companies that do heifer DNA testing. They report several traits of interest to ranchers, such as calving ease, weaning weights, mature weights, milk, marbling, ribeye area, etc. They can also combine all of these traits, weight them for their financial impact, and give the comparison between this heifer and how she ranks on a scale of 0 to 100 with the national herd of the comparable breed. If the producer analyzes the DNA results, it becomes easy to identify which heifers are not going to produce good calves.

DNA testing is not as expensive as most ranchers think. The testing currently costs about $28 per heifer sample. To begin the DNA procedure, a rancher should separate from the annual calf crop those heifers that look like they may be a good replacement heifer using their physical appearance of bone and muscular confirmation. Then a sample is taken from each calf for DNA testing. The test results will quickly identify the 20 to 30 percent of calves that should be sold and not selected as replacement heifers.

As an example, if the rancher estimates they need to replace 40 cows in a given year, they will collect perhaps 60 heifer samples from their better-looking heifer crop and have samples from those heifers sent in for DNA testing. When the DNA results come back in about three weeks, they can choose the 40 most desirable heifers to meet their herd goal. If the herd goal was to increase the marbling and calving ease, then they could select the heifers with high values for those traits, MB and CE. This would cost the rancher about $1,680 to test 60 calves. Although that is not cheap, it may result in a greater financial return when compared to the cost of a bull that may have the same impact toward the herd goal.

Taking a DNA sample is much simpler than most ranchers think. Merck-Allflex sells an excellent tool with a one-time cost of about $50 that makes taking a DNA sample easier than putting a tag in a calf's ear. The tool, like a leather punch, punches a small sample from the calf's ear and places it directly into a capsule that can be sent to the DNA company. The sample capsule used for receiving the ear sample costs about $3 each. The capsule has a barcode (with numbers) and a place for the rancher to write the heifer tag number. The total cost to the ranch for each heifer tested would be $31. This takes all the guessing and uncertainty out of replacement heifer selection.

Herd Improvement by Culling Mother Cows

Most ranchers cull cows because of their age—when they don't think she can have and raise another calf. Other ranchers may just cull their mother cows when they reach a fixed age—either to avoid having an older cow in the herd that is more likely to have a poorer calf or maybe because they can get a higher price by selling a healthy, younger cow.

However, there may be other reasons for culling a mother cow, such as producing poorer calves. We can sometimes observe that some cows just consistently produce poor calves. Of course, it would be better if one had a better measurement of producing a poorer calf.

One way of selecting for poor calf production is by reusing DNA test results from tests for replacement heifer selection as described in the above section. After you select the replacement heifers that have the best DNA traits, you could sell the mothers whose heifers showed below acceptable trait levels.

Another way to cull mother cows is to look at the quality ratings of their calves that are slaughtered for beef. In my case, I own my calves all the way through the feedlot until they are sold to a packing plant. At the time they are sold to a packing plant, I request that I get back the individual carcass gradings on each calf, such as Prime, Choice, or Select. The data collected on each carcass is presented in detail. There are actually 20 grades of Prime, 30 grades of Choice, and 10 grades of Select. The data will tell where each calf ranks within a particular grade. Even if you sell calves to the feedlot, you can request individual carcass data on your calves once they are slaughtered. The carcass data costs about $5 per head.

As my herd rated higher in quality, I looked at data from the slaughtered fattened calf's carcass grading, and any calf whose carcass was graded Select, I would find the mother of the calf and sell her.

Herd Improvement by Increasing the Quality of the Beef Produced

The term *quality* is used to define many different characteristics, but with cattle, and especially beef, it is used to mean the amount of marbling in the carcasses from the calves in the herd. About 15 to 20 years ago, packing plants began paying more for beef cattle that had higher-quality carcasses. This has been referred to as *buying or selling on the grid*, where the grid is a value-based formula that determines the amount paid for a carcass. It is indeed a system that pays a different price for hung carcasses that grade higher for marbling, and it is based on the live weight of the fattened calf entering the packing plant.

At a packing plant, each fattened calf is slaughtered, reduced to two sides of beef, and cooled in a refrigerated room overnight. The next day, the carcass is cut between the 12th and 13th ribs, and the degree of marbling in the ribeye is measured—often by a computerized camera. That level of marbling is assigned to the entire side of meat. More marbling in a carcass results in a larger payment based on the live weight of the calf when it entered the plant. Thus, a different price is paid for each animal processed. Higher marbling in a carcass increases the total price, perhaps by a $200 to $300 bonus per carcass. Higher-quality carcasses coming out of the feedlot may make the total difference between the ranch being profitable or not.

Herd Improvement by Increasing Feed Conversion Efficiency

One cattle trait that is not often talked about in cattle genetics and cattle improvement is the variation between individual cattle in their ability to convert feed to maintain themselves and/or to grow. This is obviously a very important trait, but so far, it has been difficult to measure, predict, or find genetic markers to identify why there is such a large variation between similar cattle. About 20 years ago, a system was developed to measure this characteristic, but it was time-consuming and required very expensive equipment. The first system made was called the *Grow Safe* system.

A characteristic that we ranchers can relate to is the feed-to-gain ratio (F:G). This ration has always been important in feedlots where their business is to feed yearlings to fatten them for the packing plants. They had noticed that the F:G varied between animal pens but was averaged about 7 pounds of feed per pound of weight gained. A pen usually had about 100 head in it. However, after an accurate measurement was developed for individual animals, they found that individual F:G ratios varied from about 5 up to about 10. The article about these tests that I first remember reading described a seedstock producer in South Dakota that used a Grow Safe system to measure the different F:G ratios for more than 100 uniform-looking yearling bulls he had raised and were going to his annual sale. The worst bull required 12 pounds of hay per pound of gain, and the best bull required only 4 pounds of hay to put on a pound of weight. That is a factor of three difference within a uniform-looking set of bulls that had all been raised together on the same ranch.

The unbiased way to measure, report, and select for their feed conversion efficiency (FCE) trait is their measured residual feed intake (RFI). The RFI is the measurement of the difference between an animal's actual feed intake over time and their predicted feed intake for maintenance and growth, considering their body weight and weight gain. If one used the F:G ratio for a selection measure, it would favor larger animals. The RFI measurement averages all cattle in the group being tested and sets the average at zero and thus reports the pounds more or less than the average that it takes for an individual animal to gain a pound. The more desirable cattle (higher efficiency) will have negative RFI numbers, and the more inefficient cattle will have positive values. Reported RFI numbers typically range between plus 10 and minus 10. However, since individuals in a measurement group are measured relative to others within the same measurement group, it is difficult to compare bulls from different measurement groups. The best way to ensure that one chooses a highly efficient bull is to go to a bull producer that has been selecting within their own herd for FCE and then select a highly efficient bull from that herd with a large negative RFI.

When I was making decisions to select bulls, I used the bull's numbers for feed efficiency. More recently, genomic markers have been identified to generate EPDs for feed efficiency, which improves the selection process for bulls. The RFI efficiency characteristic is known to be moderately heritable (0.18 to 0.49), and if one selects bulls that have been measured to have higher FCE, then you can make feed efficiency improvements in your herd.

It appears that the FCE of your own herd and that of the US herd could be increased by 10, 30, 50 percent, or more. This means that your costs per year for feed or forage required could be reduced by those same amounts, as well as the comparable amount of manure, methane, etc. That would be a huge improvement for our herds. However, FCE is hard to use in practice as a selection criterion.

Herd Improvement by Using Artificial Insemination (AI)

Bull semen selection and artificial insemination are the premier processes for improving the quality of the cattle. See Chapter 13,

"Cattle Operations: Artificial Insemination," for the discussion of artificial insemination.

Herd Improvement by Herd Management

Many western commercial ranchers operate by leaving bulls with the cows all year and periodically removing calves for processing, such as branding, vaccinating, or weaning. This results in calves being born all year long and thus sold at different times or discounted by buyers because of size variation. To reduce labor on the ranch and increase the price paid for calves, the rancher should consider operating the ranch with cows on the same defined schedule for all cows—within that group. This means having a fixed time period of the year for breeding, calving, weaning, and selling calves. This schedule reduces overall labor costs and produces a more uniform calf crop—desired by calf buyers and feedlots.

Other Considerations for Herd Improvement

Retained ownership of calves through the feedlot may result in a higher price return per calf. Traditionally, calves have been sold at the sale barn or directly to a calf buyer. These buyers then typically sell to a *stocker operation* that usually puts calves on grass that allows calves to grow up to the size, perhaps 800 pounds, that feedlots like to buy. If a rancher keeps the weaned calves in an available pasture and supplements their feed somewhat for four to six weeks, they can be grown out more and are then well weaned from their mothers. This is called *backgrounding* the calves. Calves can usually be sold to a calf buyer or sent directly to the feedlot at a much higher price. If a rancher retains ownership through the feedlot and sells them to a packing plant, they can recover all the profits that otherwise would be made by the middlemen. Retaining ownership is more advantageous for ranchers if their calves grade higher and are very healthy.

The *uniformity* of calves can make a big difference to a buyer. If your calves are all about the same size and have a uniform color, then they can be expected to all feed about the same way in a feedlot and be fed and managed uniformly. Also, calves that are mostly black, for example, Black Baldies, bring a better price since their black color implies that they have Angus blood in them and perhaps have higher quality meat.

Crossbreeding is very beneficial for commercial ranchers. Most favor just two breeds that complement each other. Many crossbreeds are with the Angus to incorporate the Angus carcass benefits. Examples are SimAngus (Simmental X Angus), Brangus (Brahman X Angus), Balancer (Gelbvieh X Angus), and Black Baldies (Hereford X Angus). The crossbreeds bring increases in calf survival, calf weights and growth, and greater cow longevity.

A *complete vaccination* program not only keeps cows and calves healthy on the ranch but, more importantly, makes the calves much more attractive to the calf buyers who pay more for those calves. If the health program and calf history are verified by an independent third party, that adds considerably to the price offered for the calves. For example, a common description of calves for sale is VAC45, which means that these calves have the specified vaccinations and have been backgrounded for 45 days.

Age and Source Verified (A&S Verified) calves are worth more. This term is used for a group of calves that have been inspected and verified by a third-party company as to their age, born between two specific dates, and that have been raised on a particular ranch. Due to the increase in consumer awareness and concern about "where their meat comes from," consumers will pay more for the meat. Packing plants pay a premium for A&S Verified calves, and thus the feedlots and the cattle buyers pay more.

The *Beef Quality Assurance* (BQA) guidelines should be followed because they benefit the ranch and because they make the calves more valuable to buyers. All people processing cattle on the ranch should be BQA certified. Humane treatment of all ranch animals makes sense (plus dollars and cents). I sell all cattle that are high-headed—cattle that are nervous, flighty, or those that run away from a person on horseback—even if they have good calves.

Chapter 16

HERD IMPROVEMENT: MY PERSONAL EXPERIENCE

In my case, we have owned a 145-year-old cattle ranch for the last 45 years in the rough Superstition Mountains of Arizona, about 50+ miles east of Phoenix. It consists mostly of Arizona State Trust lands and shares 10 miles of our northern border with the USFS Superstition Wilderness Area. It is characterized as a high desert ranch with canyons, rocks, cacti, and bushes. We can drive into the headquarters with hay and cattle trucks, but the cattle operations are strictly by horseback. It was the first solar photovoltaic-powered (PV) farm or ranch in the world in 1979, and it is still seven miles from the nearest electric line.

I was employed near Phoenix, full-time, until retirement in 2004, becoming a full-time rancher since then. I had employed a resident caretaker-ranch manager, and I spent weekends, holidays, and vacations at the ranch. I also depended on family members and friends to help work and move cattle. I operated a traditional Arizona commercial cattle ranch, meaning a Brahman-Cross mixed breed herd, and I ran the bulls with the cows all year. After retirement, I spent two years studying the cattle industry, cattle genetics, feedlots, packing plants, and the direction of the cattle and beef industry. In

2007, at age 70, I decided to completely change my entire herd and how I operated.

Ranch Goal

I set my new ranch goal to *maximize the price for my calves by retaining ownership through the feedlot and to qualify for all the premiums that packing plants offer.* This was a major challenge for a very rough-country Arizona rancher who had a Brahman-Cross herd and had never sent calves to a feedlot.

I learned that the packing plants pay the maximum premium for the following calf categories, so I concentrated my improvement efforts toward meeting these requirements:

- Black-hided calves, which is indicative of a high percentage of the Angus breed.
- Certified natural calves that have no antibiotics, hormones, or animal extracts in their feed.
- Source and age-verified calves that are inspected and verified by a third party.
- Unbranded calves or calves that have no scars on the hide, such as freeze-branded calves.
- High marbling carcasses grade higher and pay the best premium—Prime grade is preferred.

Changes Needed to Meet My Goal

This approach required major changes in my herd and in my operations. I thought that the first step in planning how to make changes was to send a truckload of calves to a feedlot, retain ownership, learn more about how the feedlots and packing plants work, and get individual carcass data back on each calf before deciding on my herd goal.

The effortless way to change quickly was to buy the high-quality heifers, cows, and bulls that had the characteristics desired to meet my ranch goal. The main reason this was not possible, besides the cost involved, was because experience had shown that mature animals brought into our rough terrain country nearly always died. Their feet got too sore from walking on the rocks, plus they did not know what to eat.

Therefore, I had to change my whole herd by the slow process of bull selection. I experimented with artificial insemination (AI) to see if my cows would respond and if my working facilities were equipped for AI processing—I had only a squeeze chute. The very first time I tried AI, I had a 45% pregnancy success rate, which was good enough to continue the experiment. So I decided to raise my own superior replacement heifers by extensive use of AI. With AI, I did not have to keep live bulls with superior semen on my ranch. I would just use their semen, and their calves would be raised by my native cows. The calves would then be well acclimated to my rocky country and sparse feed conditions.

Moving toward my herd goal, I decided for both my live and AI bull selections I would focus on marbling to move toward higher-grading carcasses. I'd also focus on feed conversion efficiency to minimize costs in the feedlot as well as to better maintain my future cows that would produce high-quality calves on the sparse feed in my pastures.

There were no selection tools developed yet for selecting bulls for feed efficiency, but the trait is considered moderately heritable. Thus, I buy bulls or use semen from bulls only if they have been feed efficiency tested and have measured high efficiencies (FCE) or a low residual feed intake efficiency number (RFI). As a result, I buy bulls only from the Midland Bull Test Center, the Leachman Cattle Company, or Bob Prosser's Bar T Bar Ranch of Arizona. It is easy to find bulls that have been GE-EPD and/or IMF tested for high marbling.

Besides requirements for bulls to have high carcass quality and feed efficiency, the new herd goal required operational changes at the ranch. This included both the increase of calf carcass quality by bull selection and the culling out of the bottom end of the cows that are producing poor calves. The average quality of the calves is improved more by eliminating the bottom end of the herd than by increasing the production of higher-quality calves by bull selection.

Adopted Changes

The changes required better record keeping as well as implementing the following:

- Rode regularly during the calving period to rope new calves, put a calf-size ranch tag on each newborn, and record the mother's tag number and the calf's date of birth.

- Installed EID tags on calves, in addition to the ranch tags at branding time.

- Changed to freeze branding that leaves no scars on the hide.

- Made sure I gave all shots that the feedlots wanted and still met the certified natural requirement.

- Had an annual inspection at the ranch to have third-party verification for an all-natural operation and be age- and source-verified.

- Took all calves off the cows in the fall during pregnancy checking and kept them at the ranch for backgrounding.

- Backgrounded calves for six weeks at the ranch before shipping to the feedlot. Administered all needed booster vaccinations.

- Kept some heifer calves for replacements and sent all other calves to the feedlot in December.

- Specified that the feedlot require the packing plants to report back to me the complete carcass data for each calf slaughtered.

- Analyzed our records to find the mother of any calf grading only Select, not Choice or Prime, and sold her, thus eliminating her from the herd.

- Went to a three-month calving season, mostly so that the higher-quality live bulls I buy are only with the cows in the least rough pasture for the breeding season. This procedure also works well for feedlot shipping schedules.

- Usually brought the cattle in at the beginning of June, and the bulls were placed in a special bull pasture.

- Sold open cows. When the cows and calves were brought in from the summer pasture, the calves were taken off, the cows were pregnancy checked, and if open, they were sold. This culling will eventually improve the fertility of the herd.

- After pregnancy checking, the cows are moved to the fall and winter pastures for calving.

- Brought cows and calves in for branding in March; the cows are all synchronized for 10 days, 95% of them are AI-ed, and then they are turned out with the bulls for cleanup. The AI results in about 55% of the cows being bred.

- Started DNA testing some heifer calves about 2007. I selected the heifer calves that physically looked like prospective replacements and took DNA samples at $2 to $3 per sample. I sent in about 150% of the number of replacements that I needed, and after I received the DNA results, I selected the top two-thirds for replacements. The testing cost was $28 per calf tested.

The costs of these changes in operations are mostly in labor time. The biggest cost is to feed the calves for six weeks. However, this replaces some of the costs I would have had in the feedlot. The AI cost per live calf is about the same as a naturally sired calf. Of course, I need only about half the live bulls, since half of the calves are from AI. This reduction in live bulls also allows me to run more cows on the ranch.

Because of the operations described above, all my cows were born and raised on the ranch and obviously acclimated to our harsh living and feed conditions.

In addition to my two main herd goals focusing on bull selection for marbling and FCE, I set a secondary goal to aim for a herd that consisted of about 75% Angus and 25% Gelbvieh breeds called Balancers. Balancers have the high marbling characteristics of the Angus, the high mothering and adaptability of the Gelbvieh, and the hardiness of the crossbred. For cows that had an Angus sire, I AI-ed them to a Balancer bull that had about 50+% Angus. And for a Balancer-sired cow, I AI-ed them to an Angus bull.

For more information on goals, see Appendix A, "How to Implement Your Herd Goal."

My Mentors

I had two major mentors in my quest to improve my herd. One was the largest seedstock producer in Arizona—Bob Prosser of the Bar T Bar Ranch. Bob has an annual bull sale of 200+ registered bulls and measures more data on them than anyone I know. He also runs a very large commercial herd, thus having a commercial rancher's perspective. I have bought more bulls from him than any other source during the last 10 years. Bob has extensive experience with crossbreeding cattle and has narrowed his focus to mostly producing registered Balancer bulls. The Balancer is an Angus and Gelbvieh crossbreed of 25% and 75%, respectively. Bob bought the first Grow-Safe equipment in Arizona and has been reporting the RFI measurements on all 200 bulls sold each year for about ten-plus years. He focuses his registered herd on low RFI, low birth weight, and well-balanced EPDs across the spectrum. Since he reports measured RFI, GE-EPDs, and measured data including IMF, feed intake, and ADG on so many bulls, I can usually find the types of bulls I desire to meet my herd improvement goal. I have also partnered with him on buying outstanding Balancer and Angus bulls from other sources. In those cases, he maintains the live bull, and I get as much semen as I want for AI-ing at my ranch. The fact that the live bulls I buy from him are grown in Arizona on sparse feed conditions allows them to easily adjust to my rough country.

The second "mentor" I had is the entire crew at the Certified Angus Beef (CAB) organization. As I was defining my herd goal to raise high-quality calves, I found that CAB required these types of beef carcasses for the beef products that they certify. They were very helpful in directing my education in the grading process used in packing plants. In order to be certified for CAB approval, meat is required to be graded Premium Choice or Prime. Thus, beef qualified to be CAB is indeed very highly marbled, resulting in an exceptionally satisfying eating experience—juicy and tender steaks.

Selecting a Feedyard

To begin my work with feedlots, I sent a truckload of calves to a small, custom feedlot in the Texas Panhandle. I visited the feedlot before sending calves to them and a couple of times during their feeding period. I requested detailed carcass data from the packing plant for each carcass, which cost about $5 extra per carcass. The data indicated that they were about 50% of both Select and Choice—the national average at the time. Later, when I learned how to interpret the carcass data, I found that the carcasses graded Choice were in the bottom group of the 30 sub-grades of Choice.

A few years later, that small feedlot was sold to a larger feedlot. In talking with the people at CAB, they suggested that I should consider feeding at Dale Moore's Cattleman's Choice Feedyard that the CAB had given the Small Feedlot of the Year Award the year before. Tony and I visited with Dale in Oklahoma, and I decided to work with him. At the time, the Cattleman's Choice Feedyard was a small, family-owned feedlot and fed about 7,000 cattle. I have fed there for the last 12 or so years. It has now grown to their maximum size of 10,000 cattle. They are a *program-cattle-only* feedlot. The owner does the hedging of the calves and marketing to the packing plants.

Herd Goal for Premiums Achieved

The changes I made in my operations to approach my herd goal have been gradually improving the carcass quality of the calves for maximizing my premiums at the packing plant. The marbling trait is moderately heritable, so I have been consistently monitoring my quality grading levels at the packing plant over the years. There are 30 sub-grades in the Choice grade, so I was able to observe the upward movement of my quality grades. For the last couple of years, my overall carcass grades have been 100% Choice or better, 50% Prime, and 40%+ Premium Choice. Over time, when the carcass data indicated just a very few calves were grading Select, I went back to my herd and sold all the mother cows that produced those Select-graded calves. This culling process sped up my transition to a higher-quality herd.

I am getting close to meeting my herd goal for quality premiums, but I can continue to increase the percent of prime-quality carcasses. The higher feed efficiency is much harder to make progress on and is still a desired, longer-term goal.

Powder Spring Canyon. Chuck is closing the pipe welder to weld the two ends of the poly pipe together. The gasoline generator supplies 110 volts AC to the welder heater. Photo by Jack Carlson, February 2006.

Powder Spring Canyon. A closeup of the welder with the two poly pipes clamped together while the weld is being made. Photo by Jack Carlson, February 2006.

Powder Spring Canyon. A closeup of Chuck closing the pipe welder to push the poly pipe ends against the heater block, which will melt the ends of the pipe. When the heater is removed, the pipes are pushed together, and the melted ends are joined to form the weld. The gasoline generator is supplying 110 volts AC to the welder heater. Photo by Jack Carlson, February 2006.

Powder Spring Canyon. A closeup of the completed weld looks like a bead or grommet around the pipe. This weld is much stronger than a connection made with an insert or clamped coupling. Photo by Jack Carlson, February 2006.

Powder Spring Canyon. Chuck, right, and Howard, left, are routing the poly pipe down canyon to supply water to additional drinkers. The water originates at Powder Spring higher up in the canyon. Photo by Jack Carlson, February 2006.

Chuck, right, and Howard, left, are moving the welding gear farther down Lower Powder Spring Canyon to connect more pipe. Tonopah is the packhorse. The long tools, such as a shovel and a digging bar, are usually strapped on the sawbucks of the packsaddle. Howard's horse, Niles, is on the left. Photo by Jack Carlson, February 2006.

Chapter 17

SOLAR INSTALLATIONS AT THE RANCHES

Since I was one of the world's pioneers in the development of photovoltaic (PV) cells for terrestrial use, it seems fitting that I should install PV arrays at our ranch.

In 1973, the National Science Foundation funded a $200,000 grant to me at ASU for a joint project we had proposed to study solar energy for terrestrial applications by using solar concentrators on solar cells. ASU was the prime contractor, with about half of the money going to a subcontract with Spectrolab Corporation in California, one of two companies that made solar cells for space satellites, to study how to make cells that would perform better with higher concentrations of sunlight. This led to more than 20 years of continuous funding for me and my ASU students.

In 1973, after a workshop on space photovoltaic systems held at Cherry Hill, New Jersey, some attendees from ESSO Research Labs decided to try making terrestrial solar modules directly from silicon wafers mounted on circuit boards. They sent their modules to NASA-Lewis for terrestrial environment testing, and after a year they were put in storage. My friends at Lewis asked me if I could use them at my ranch, and I said yes. So they sent me my first modules of solar panels. My ranch thus became the first one in the world to be powered by photovoltaics.

Soon new companies making photovoltaics for terrestrial applications started sending panels to me for outdoor testing. This trend of increasing power capacity at the headquarters continued to the point where we had air conditioning in the ranch house and in the manager's house. The entire ranch soon became solar-powered.

At the headquarters, the solar panels are connected to storage batteries and an inverter that supplies 120 or 240 volts AC to the buildings, the branding area, and the house well. For the remote wells in the pastures and one spring pump, the solar cells are connected directly and supply 12 volts DC to the pumps. Those pumps run only when there is adequate sunlight since they do not have storage batteries.

The following pictures show some of the solar-cell installations:

Howard and Chuck are pulling the submersible pump at the Upper Corral Well. The pump operates only when there is adequate sunlight and if the water level is high enough. The solar panels to the left provide electricity to the pump at the bottom of the well casing. Photo by Jack Carlson, December 2005.

Solar Installations at the Ranches

A view of the Lower Well windmill, solar panels, and concrete drinker. The windmill is not in operation. The solar panel supplies 12 volts DC to directly power a water pump for the drinker. No backup batteries are included in the installation, so water is pumped only when there is adequate sunlight. Photo by Jack Carlson, August 2005.

Howard and Chuck just pulled the pump from the well at the Upper Corral. The plan was to replace the old pump with a Grundfos submersible pump and install a new power controller box. Chuck checked the new pump by powering it with 110 volts AC from the portable generator, which worked fine. The solar cells supply a lower voltage so the pump moves water more slowly, and that is fine because it matches the inflow rate of the well better. Photo by Jack Carlson, December 2005.

Solar Installations at the Ranches

View to the northeast of the headquarters area showing, from left, the older white-colored manager's house that was removed in 2018, solar panel array, solar electronics barn, bunkhouse, corral, and hay barn. Photo by Jack Carlson, October 2007.

A view of the ranch headquarters from the south. Note the new ranch manager's home on the left with the south-facing roof covered with solar modules. Also, two tracking solar arrays can be seen to the south of the home. This system provides both 110-volt AC and 220-volt AC to all buildings at the headquarters. It powers air conditioning and heating for the manager's home. An equipment room for the solar electronics was built on the end of the manager's house and is cooled by an air conditioner that, of course, is powered by the equipment it is cooling. The bunkhouse is in the center of the photo, and to the right of the water tower is the tack room. Photo by Chuck Backus, 2016.

A view of the solar panel array at the ranch headquarters. The small, white, shingle-roofed barn is the original structure that held the rechargeable batteries and electronics for the conversion from DC to 120 and 240 volts AC for the headquarters buildings and house well. The barn has been replaced by an air-conditioned room attached to the manager's house. Photo by Jack Carlson, October 2007.

A view of the solar array at the Lower Tule Spring. The solar cells supply 12-volt DC to a water pump that is located at the nearby Lower Tule spring box. The electric cable is housed in 1-inch poly pipe to protect it from the elements and critters. Photo by Jack Carlson, May 2010.

Chapter 18

Continuing Education for Commercial Ranchers

Since I had a personal background as a university professor, a cattle rancher, and a member of the Board of the Arizona Cattle Industry Research and Education Foundation, I thought that it was appropriate that I offer continuing education for commercial cattle ranchers. I thought that ranchers needed to know more about the entire cattle industry and what happened to their calves after they left their respective ranches. Thus, in February of 2014, I offered a one-day workshop for commercial Arizona ranchers on "Increasing the Health and Wealth of Our Calves." About 65 commercial Arizona ranchers attended. I solicited enough company sponsors to cover the costs of the workshop and lunch for all attendees. The commercial ranchers were so impressed with the presentations and the workshop that they requested that I do additional educational programs every couple of years.

For this first workshop titled "Increasing the Health and Wealth of Our Caves," I invited national speakers, mostly from the Midwest, to speak on the following topics:

- The kinds of calves the feedlot owners wanted.
- What the feedlots wanted relative to the medications given.

- What packing plant owners wanted and expected for incoming calves.

- The importance of using genetics to make their calves worth more.

- The importance of EPDs, ultrasound, and feed efficiency for calves.

- The importance of buying the correct bull to produce the calves desired.

The lessons learned by ranchers at this workshop were: You cannot tell quality from looking at a calf; it is hard to compete with the beef industry's sophistication and efficiency; the various sectors of the beef industry make marginally small profits; and some in the beef industry make money only because of the high volumes they process.

In February 2016, at a larger hotel—the Four Points Hotel, by Marriott in Ahwatukee—we had another workshop titled "Maximizing the Value of Our Arizona Ranch Calves: Mostly by Selectively Choosing Better Bulls." At this second workshop, we had about 95 attendees. The workshop focused primarily on how to make the calves worth more by improving their health and the quality of their carcasses. Thus, there were a lot of presentations about expected progeny differences (EPDs) and keeping calves healthy. Discussions about marbling in calves brought out the fact that higher marbling could bring a higher price. And there seemed to be more differentiation in beef products because customers were becoming more aware of higher-quality beef. This may have been because products like Certified Angus Beef (CAB) and other quality beef products were more readily available.

The next workshop was held on February 15 and 16, 2019. It was called "A Cow-Calf Symposium for Southwestern Ranchers." As the workshops built somewhat of a reputation, this one had an attendance of 120 people and was advertised for all Southwestern US ranchers.

There were more nationally known speakers, such as Mark McCully from Certified Angus Beef, Bob Prosser from the Bar-T-Bar Ranch, Matt Spangler from the University of Nebraska, Leo McDonnell from the Midland Bull Test Center, Bob Weaber from Kansas State University, Lee Leachman from Leachman Cattle Company, John Patterson from Neogen/GeneSeek Company, Jim Johnson from Zoetis Company, Jim Gibb from GeneSeek, Matt Herrington from STgenetics, and Chuck Backus, the symposium organizer.

There were no workshops in 2020 and 2021 because of the COVID-19 pandemic. I was retiring, and my last symposium was called "A Herd Improvement and Bull Selection Symposium," which was held on February 18 and 19, 2022, at the Mesa Hilton. It was advertised for all Western commercial cattle ranchers. We had about the same number of presenters as the last symposium, but it attracted an attendance of 180 people. We had a wide range of topics on cattle quality, bull selection, and general herd improvement. The topics emphasized the balance of characteristics that made for a better eating experience and for enjoying a good beef meal. Thanks to our sponsors, none of the workshops or symposia required an attendance fee, and all meals were free.

The following is a copy of the program for the 2022 "Herd Improvement and Bull Selection" symposium.

Herd Improvement and Bull Selection
A Cow-Calf Symposium for Commercial Western Ranchers
Organized by the Arizona Cattle Industry Research
and Education Foundation—ACIF

Friday, February 18, 2022
12:30 p.m.—**Introduction and Overview**
    (Chuck Backus, Arizona Commercial Rancher, ACIF Trustee, and Symposium organizer)

1:00 p.m.—**National Beef Cattle Status/Trends**—the Importance for Commercial Ranchers
(Paul Dykstra, Certified Angus Beef)
1:30 p.m.—**Historical Perspective of Cattle Genetics**—A Timeline of Understanding and Using Techniques for Herd Improvement (Genetics, DNA, EPDs, GE-EPDs, etc.)
(Dr. Matt Spangler, Professor at the University of Nebraska)
2:15 p.m.—**Current and Future Genetic Research and Projections**
(Dr. Alison Van Eenennaam, Professor at UC Davis)
3:00 p.m.—Break
3:15 p.m.—**Other Methods of Herd Improvement**—Crossbreeding, etc.
(Dr. Robert Weaber, Professor at Kansas State University, President of BIF)
4:00 p.m.—**Panel on Genetics and Herd Improvement**
(Spangler, Van Eenennaam, Weaber, Dykstra, and Backus)
5:00 p.m.—Cash cocktail hour
6:00 p.m.—Banquet
6:30 p.m.—**A Simpler Selection Technique**
(Lee Leachman, CEO/Owner of Leachman Cattle Company, Colorado)

Saturday, February 19, 2022
8:00 a.m.—**Setting Goals for Herd Improvement**
(Chuck Backus)
8:30 a.m.—**Alternatives for Selling Better Ranch Calves**
(Paul Dykstra, Producer Relations, Certified Angus Beef)
9:00 a.m.—**Replacement Heifer Selection and Techniques Available**
(Dr. Kelli Retallick, President, Angus Genetics Inc.)
9:45 a.m.—**The Importance and Simplicity of Artificial Insemination**
(Roger Wann, District Manager, American Breeding Service (ABS))
10:15 a.m.—**Herd Improvement by Increase of Feed Efficiency**—What It Is, Measuring It, and Saving 20% in Feed Costs
(Leo McDonnel, Midland BTC & Commercial/Seedstock Rancher, MT)

10:45 a.m.—Break

11:00 a.m.—**The Terminology in Selection**—Understanding and Using Bull Sale Catalogs, Using EPDs, Measurements & Techniques for Selecting Ranch Bulls

    (Bob Prosser, owner of Bar T Bar Ranch, seedstock and commercial producer, largest bull sale in Arizona)

11:30 a.m.—**Panel Discussion**—All Speakers

    (Dykstra, Retallick, Wand, McDonnel, Prosser, and Leachman)

12:30 p.m.—**Conclude Symposium**—Thanks for coming.

I sincerely enjoyed this last symposium and all the ones that preceded it. I have served as the president of the Arizona Cattle Industry Research and Educational Foundation for three years, but no activity was more rewarding to me as an individual than these four different symposia I have organized and run. Of course, my wife Judy was just as involved as I was.

Chapter 19

The Transfer of Ranch Ownership

The transfer of ranch ownership is not discussed in most books on cattle ranching, but it is ultimately encountered by all ranch owners. Often this is not a problem because one of the descendants is ready to assume the responsibility of the ranch ownership. However, if there is more than one descendant, it becomes a problem of how to be fair to all descendants.

In our case, first we sold the Northern Ranch to one of the adjacent ranchers for cash. That was an easy one. However, our youngest daughter, Amy, and her husband, Mike Doyle, were interested in buying the Superstition Ranch, and Tony and Beth were not. We first went to the Farm Credit Corporation, the company that had funded us to buy the ranch, and asked them to appraise the current ranch and the cows on it, and that resulted in the sale of the Superstition Ranch to Amy and Mike in 2022.

Then we addressed more of our retirement issues in the following manner: First, we sold our Gilbert home and moved into a retirement community where we did not have to own the home, but we could modify it to our choosing, but not have to buy—only a monthly rate—until we both pass. Then, we loaned money to our other two children to pay off their home loans, but at an interest rate much lower than their bank loans—but at a rate that our tax man said was acceptable for our tax purposes. That way, all our kids pay monthly payments to us until we both pass.

Since Judy and I both worked during our lives, we both have a nice retirement income. We have money from the sale of our Northern Ranch for other investments, and we are financially secure for the remainder of our lives.

These young heifers are going back to the Heifer Pasture after they were processed to start the artificial insemination synchronization protocol. They will be brought back to the headquarters in seven days to receive the artificial insemination. The peak on the horizon is named Miner's Needle, and it is within the boundaries of the Superstition Wilderness. Photo by Chuck Backus.

Appendix A

How to Implement Your Herd Goal

Throughout this book I have stated my herd improvement goal in one long sentence. This is helpful to quickly convey my improvement objectives to the ranch personnel, friends, ranch volunteers, and the feedyards. But to implement the goal, I needed to drill down into the details and set specific actions that I could take.

An example of a one-sentence goal mentioned in this book is to *produce calves that would grade Prime or Certified Angus Beef quality cattle, both steers and heifers, coming out of the packing plants.*

A revised goal was to *maximize the price for my calves by retaining ownership through the feedlot and to qualify for all the premiums that packing plants offer.*

The following is an example of a detailed action plan that I used to implement the February 9, 2012, Quarter Circle U Ranch herd goal. This early plan shows that there were many areas for improvement when I decided to begin this experiment. In contrast, since we have achieved our early objectives, our most recent plan focuses mostly on obtaining the packing plant premiums.

Operational and Marketing Goal for the Herd

- Expand to 450 Angus-Cross mother cows.
- Raise more than 375 healthy calves per year.
- Extensively use AI and buy very high-quality bulls.
- Raise our own replacement heifers.
- Send all steers and non-replacement heifers to the feedlot, retain ownership, and sell to markets that pay premiums.

Specific Goal for the Herd

- Calve in January and February with a few calving in March.
- Send all calves to irrigated pastures in late November for both heifer development and feedlot preconditioning.
- AI selected heifers (low birth weight sires), do a short-term preg test, AI open heifers again, preg test again in May, keep all pregnant heifers, and sell open heifers.
- Send all steers and some heifers to the feedlot in February and March.
- Try to maintain a 95% pregnancy rate in the fall preg tests.

Implications for AI Sire Selection for Cows for Arid and Rough-Ranch Conditions

- Use only Angus sires with a good balance of traits.
- Select high values for, in priority order—FCE, marbling, $P, $G, and YW.

Appendix A: How to Implement Your Herd Goal

- Select above average values for CEM, REA, CW, CED, DOC, $EN, SC, and HP.
- Select average or below values for BW, Milk, and Fat.

Implications of Goals for Buying Bulls for Natural Service

- Buy Angus and Balancers Bulls at the Bar T Bar sale that provide EPDs, ultrasound and feed conversion measurements, and DNA tests.
- Buy bulls with high values, in priority order—FCE, IMF, ADG, and REA.
- Otherwise use the same selection priorities as in AI sire selection above.

Abbreviations for EPDs

$B	Beef Value
$Cow	Profit Value
$EN	Cow Energy Value
$F	Feedlot Value
$G	Grid Value
$M	Maternal Weaned Calf Value
$P	Profit Value
$W	Weaned Calf Value
ADG	Average Daily Gain
BW	Birth Weight
CE	Calving Ease
CED	Calving Ease Direct

CEM	Calving Ease Maternal
CW	Carcass Weight
DOC	Docility
Fat	Fat Thickness
FCE	Feed Conversion Efficiency
F:G	Feed to Grain Ratio
HP	Heifer Pregnancy
IMF	Intermuscular Fat
Marb	Marbling
Milk	Maternal Milk
REA	Rib Eye Area
RFI	Residual Feed Intake
SC	Scrotal Circumference
STY	Stayability
YW	Yearling Weight

Appendix B

Expected Progeny Differences

By Matthew L. Spangler, Professor at University of Nebraska-Lincoln

Excerpt from the presentation at the February 2022 symposium "Herd Improvement and Bull Selection: A Cow-Calf Symposium for Commercial Western Ranchers"—organized by the Arizona Cattle Industry Research and Education Foundation—ACIF.

Dr. Spangler's presentation was titled "Historical Perspective of Cattle Genetics—A Timeline of Understanding and Using Techniques for Herd Improvement (Genetics, DNA, EPDs, GE-EPDs, etc.)." This paper is titled "Expected Progeny Differences."

Introduction

It is impossible to visually determine the genetic potential of an animal as a parent for traits that are controlled by numerous genetic variants, as is the case for fertility, growth, carcass merit, and other trait complexes of economic importance. Consequently, predictions of genetic merit have evolved over the last several decades and now include phenotypic information, pedigree information, and more recently genomic information. These predictions are called Expected Progeny Differences (EPD) and have been proven to be the most reliable tool to generate change from selection.

What Are Expected Progeny Differences?

Expected Progeny Differences are predictions of genetic merit of an individual as a parent. As the name would imply, they are predictions of the differences in individuals' offspring performance. Historically, most beef breed associations conducted a genetic evaluation twice annually, meaning that EPD were updated twice a year. This schedule was due to the fact that new data were generally available twice a year, to correspond with two general calving seasons (spring and fall). However, with the advent of genomic information, new data are continually available, and producers wish to see the changes in EPD that result from the new data. This has necessitated weekly genetic evaluations, and thus updated EPD are available on a weekly basis for the majority of beef cattle breeds. In other words, more frequent genetic evaluations mean more current predictions of the genetic merit of animals.

How Do You Use EPD?

Expected Progeny Differences are tools designed to compare animals based on their genetic potential as parents and to make directional change for a particular trait. Simply knowing an animal's EPD for a given trait has no meaning without something to compare it to. This comparison can be between animals or between an animal and a point of reference, such as the average of a particular breed. Breed averages are rarely 0. Rather they represent either a point in time or a set of reference animals (i.e., historic set of high accuracy sires). Knowledge of breed average is helpful in determining how an animal ranks within a given breed for a particular trait. Most breeds publish a percentile rank table which allows producers to determine how an animal ranks for a particular trait within a particular breed. Expected Progeny Differences are reported in units of the trait. For example, weight traits (e.g., birth, weaning, yearling) are reported in pounds. However, some traits are reported as percentages (e.g., heifer pregnancy, docility).

With this in mind, the interpretation of the difference in EPD between two bulls is the average difference in performance of their offspring if the bulls were mated to the same cows and the calves were reared in the same environment. The following is an example.

Example Bull A has a weaning weight EPD of 50.
Example Bull B has a weaning weight EPD of 60.

Based on this example, on average, we expect the offspring of Bull B to weigh 10 pounds more than the offspring of Bull A. This does not mean that every calf from Bull B will weigh more than every calf sired by Bull A. There will be variation in the weights of calves produced by both bulls, but with large enough groups of offspring the average difference will be reflected by the difference in sire EPD.

Calculating EPD

The actual calculation of EPD requires the use of sophisticated statistical approaches and modern computational resources. To put the task into perspective, larger breed organizations calculate EPD for approximately 12 to 20 traits for more than 10 million animals on a weekly basis. This is not a trivial task. However, the calculation of reliable EPD begins at the ranch level. Accurate phenotypes and correct accounting for management differences is the responsibility of the breeder. Although advancements in the use of artificial intelligence (AI) algorithms are beginning to penetrate genetic evaluations as a means of categorizing data in terms of quality, the fundamental responsibility will always belong to the breeder to ensure that records are accurate. Records collected at the ranch level are sent to breed organizations where they are adjusted for effects such as age of the animal, age of the animal's dam, and breed composition. These adjusted records are then used in the genetic evaluation.

The genetic evaluation itself uses a system of equations referred to as the mixed model equations (MME). This system of equations uses phenotypes of animals from across the country, and in many cases internationally, to estimate the genetic value of animals. This method requires that animals are linked through relationships, either pedigree or genomic based. Given these linkages, an animal's genetic merit is informed not only by its own phenotype but also by the phenotypic records of relatives from other herds and across time. The more closely related two individuals are, the more they contribute to the other's EPD.

What Are Accuracies?

Accuracy is the theoretical correlation between an animal's EPD and their true genetic merit and can range between 0 and 1. In the U.S. beef industry, Beef Improvement Federation (BIF) accuracy is used, which is much more conservative than "true" accuracy. Expected Progeny Differences are predictions and thus are not known with complete certainty. They are updated, and become more accurate, when additional data becomes available. For example, a young non-parent animal may have a record for their own weaning weight. If the animal becomes a parent and has offspring with recorded weaning weights, their offspring inform their EPD. This increases accuracy. Another source of data that increases accuracy is genomic data. Genomic information, in the form of SNP markers, is routinely included in the genetic evaluation of all major U.S. beef cattle breeds. This enables higher accuracy predictions, particularly for non-parent animals. One way that genomic information is used to increase accuracy is by improving the estimates of relationships between animals. Instead of relying solely on pedigree information to inform kinship, genomic data can be used to determine the relationship between animals. For example, although the expectation (pedigree) of the relationship between an individual and their grandparent is 0.25, the true relationship (genomic kinship) can range between 0 and 0.5 due to sampling of alleles inherited by different animals from their parents. By estimating relationships more accurately, EPD becomes more accurate.

Contemporary Groups

A contemporary group represents a set of animals that were given an equal opportunity to perform and shared a common environment. The foundation for a contemporary group includes animals born in the same year, season, herd, and who were treated equally. In other words, if a subset of animals is fed differently (given preferential treatment) they should become a separate contemporary group. Admittedly there is an optimization between accounting for environmental effects through contemporary groups and allowing contemporary group size to be large enough to compare animals (and parents). At the limit, a contemporary group size of one would perfectly account for the unique environmental effects experienced by the animal. However, single animal contemporary groups are not useful for genetic evaluation as the animal's genetic merit becomes completely confounded with the environmental effects.

It is critical to report data on all animals in a contemporary group. Not doing so leads to biased estimates of genetic merit. If only the heaviest 50% of calves have weaning weights reported, then the magnitude of the differences between each animal and the average of the contemporary group is shrunk, incorrectly suggesting that the animals reported are not as superior for pre-weaning growth as they actually are.

Direct vs. Maternal EPD

Some phenotypes are influenced by both the genetics of the individual (direct) and genetics of the dam (maternal). Examples include weaning weight and calving ease. The EPD for weaning weight direct is simply called weaning weight whereas the maternal EPD for weaning weight is called milk. In beef cattle, milk EPD is expressed in pounds of weaning weight due to maternal influences, principally lactation. Milk EPD can be thought of as the comparison of a bull's grand-progeny that are products of his daughters. Calving ease also has a maternal genetic component. Calving ease direct EPD represent the probability of how easily a bull's calves will be born when he is bred to heifers. Calving ease maternal EPD are a misnomer in the sense that they reflect total maternal merit. Total maternal is the sum of maternal EPD and half of the direct EPD and represents the probability of unassisted births of a bull's daughters during their first parturition. Although calving ease maternal EPD are not labeled as such, the majority of beef breed associations publish total maternal calving ease.

Multiple-trait Analysis

Many traits are genetically correlated to each other. As such, knowledge of the performance of one trait informs the genetic prediction of another, correlated trait. Growth traits are a good example. Birth, weaning, and yearling weight are all genetically correlated with each other and as a consequence are evaluated in the same multiple-trait model. This has two primary benefits. First, it enables early growth traits to inform the EPD of later growth traits before the later growth traits are observed. Secondly, it mitigates the impact of selection that has occurred earlier in life (sequential selection) on EPD. In the case of yearling weight, it is conceivable that animals with low weaning weights were culled prior to the collection of yearling weight. Accounting for this selection decision is critical to avoid bias in traits measured later in life, in this example yearling weight. Using a multiple-trait model accounts for the fact that selection occurred and some animals were culled while others were not. An important caveat is that although yearling weight EPD are reported, the actual trait analyzed is post-weaning gain. Resulting EPD for weaning weight and post weaning gain (adjusted to 160 days) are then summed and reported as yearling weight EPD. Another example of a multiple-trait model is calving ease and birth weight. Birth weight is a useful indicator of calving ease and is thus included in the same model as the economically relevant trait of calving ease. This means that resulting calving ease EPD incorporate birth weight observations, and selecting on both calving ease and birth weight EPD results in overemphasizing birth weight.

Multi-breed Analysis

In the U.S. beef industry there is a mixture of single- and multi-breed genetic evaluations. Single-breed genetic evaluations utilize data from only one breed, while multi-breed genetic evaluations utilize data from multiple breeds. Currently the largest multi-breed genetic evaluation is International Genetic Solutions (IGS). The goals of multi-breed genetic evaluations are sharing of data across breeds and the ability to report EPD across multiple breeds that are directly comparable to each other. The underpinning of a multi-breed genetic evaluation is pedigree ties across breeds and contemporary groups that include animals from more than one breed (or crossbred animals). Pedigree ties across breeds enable the sharing of data across breeds. Generally speaking, Angus and Red Angus serve as the links that tie multiple breeds together, largely due to composite programs such as Lim-Flex, Balancer, and SimAngus. Having contemporary groups that contain more than one breed enable the estimation of breed differences, which are needed to conduct a multi-breed genetic evaluation. Without this, breed differences must be obtained from external sources (e.g., U.S. Meat Animal Research Center).

Summary

Expected Progeny Differences enable genetic selection decisions for multiple traits. Core to accurate EPD are well-formed contemporary groups. Expected Progeny Differences change over time as additional information is available. These changes are more frequent with weekly genetic evaluations. Genomic data that is integrated into EPD allows accuracy of non-parent animals to increase.

Appendix C

Interpretation and Use of Expected Progeny Differences

By Matthew L. Spangler, Professor at University of Nebraska-Lincoln

Excerpt from the presentation at the February 2022 symposium "Herd Improvement and Bull Selection: A Cow-Calf Symposium for Commercial Western Ranchers"—organized by the Arizona Cattle Industry Research and Education Foundation—ACIF.

Dr. Spangler's presentation was titled "Historical Perspective of Cattle Genetics—A Timeline of Understanding and Using Techniques for Herd Improvement (Genetics, DNA, EPDs, GE-EPDs, etc.)." This paper is titled "Interpretation and Use of Expected Progeny Differences."

Introduction

Expected Progeny Differences (EPD) are the most reliable tools to generate directional change in traits. However, like all tools, they must be used correctly and require some degree of background knowledge to ensure proper use.

Breed Averages

Every breed provides breed averages for every trait with a published EPD. Breed average, as the name implies, is the average EPD for a given trait within a specific population (e.g., breed). Breed averages are rarely zero, but instead reflect a point in time or a set of historic animals. Some breeds further delineate breed average to subsets of animals, such as sires, dams, non-parent animals, and based on breed fractions (i.e., hybrids, purebreds, full bloods).

Percentile ranks

Breed averages can serve as a barometer relative to how an animal compares to other animals in a breed. Percentile ranks serve as a more refined gauge of how an animal compares to other animals in the same breed. Like breed averages (50th percentile), percentile ranks are available for every trait with an EPD. Depending on the breed association, percentile ranks may be available for sub populations (e.g., parent animals, non-parent animals, breed makeup). Percentile ranks indicate what proportion of animals have an EPD that is better or more desirable than a given value. As an example, an animal with an EPD in the 10th percentile means that 90% of the population has an EPD for that trait that is considered less desirable than the EPD of this animal. Note that depending on specific goals of a breeding program, extreme values may not be desirable and animals that have higher percentile ranks (e.g., 50th–99th percentile) may be desirable. An example percentile rank table is presented in Table 1. Assume a bull available at auction has a calving ease EPD of +13.0. Based on the values in Table 1, this bull would be in the top 40th percentile of the breed for calving ease. If the same bull had a yearling weight EPD of 111, he would be in the 50th percentile (breed average) for yearling weight.

Appendix C: Interpretation and Use...

%	CE	BW	WW	YW
1	19.0	-5.0	100	150
5	17.0	-3.0	89	140
10	16.0	-2.5	88	133
15	15.0	-1.9	85	128
20	14.5	-1.1	82	125
25	14.0	-0.7	80	122
30	13.8	-0.5	78	120
35	13.2	-0.2	77	118
40	12.7	0.1	76	115
45	12.5	0.3	74	113
50	12.2	0.5	73	111
55	11.9	0.8	72	110
60	11.6	1.1	71	107
65	11.1	1.5	69	105
70	10.6	1.9	68	103
75	10.1	2.0	67	100
80	9.5	2.6	65	97
85	9.1	2.8	63	94
90	8.0	3.1	59	90
95	7.2	3.7	57	85

Table I. Example percentile rank table for calving ease (CE), birth weight (BW), weaning weight (WW), and yearling weight (YW).

Possible Change

Possible change values allow producers to construct confidence intervals or ranges around an animal's EPD. Possible change is inversely related to accuracy; as accuracy goes up, possible change goes down. As compared to accuracy, possible change represents a more tangible tool to determine the risk associated with the possibility of an EPD deviating from the animal's true genetic merit as a parent. Most breed associations publish a possible change table. Possible change values are unique to each breed and each trait. To use a possible change table, the user needs to know the correct breed, trait, and the accuracy value associated with a particular animal's EPD.

Mechanically, possible change can be thought of as standard deviation and the EPD as a mean. Given this, the EPD ± the possible change can provide a confidence interval in which the true genetic merit is expected to be contained. Assume a bull has an EPD of 2.0 and a possible change value of 0.5. We expect his true EPD to be within the interval of 1.5 to 2.5 (EPD ± 1 × PC) 68% of the time. Likewise, we would expect his true EPD to be within the window of 1 to 3 (EPD ± 2 × PC) 95% of the time and from 0.5 to 3.5 (EPD ± 3 × PC) 99% of the time. The implementation of confidence intervals allows producers to visualize both the impact of improved accuracies but also enable selection whereby an animal attains some minimum or maximum threshold with some predetermined level of confidence. Confidence intervals can be very effective genetic risk management tools.

Economically Relevant Traits and Indicator Traits

The key questions that every farmer/rancher needs to answer are:

- What are my breeding/marketing goals?
- What traits directly impact the profitability of my enterprise?
- Are there environmental constraints that dictate the minimum, maximum or optimal level of performance that is acceptable for a given trait in my enterprise?

Once these three questions are answered, sire selection becomes much simpler. The answers to these questions inherently lead a producer to the traits that are economically relevant to their enterprise. We call these traits economically relevant traits (ERT; Golden et al., 2000). Fundamentally these are traits that are directly associated with a revenue stream or a cost. All traits that are not ERTs are indicator traits, or a trait that is genetically correlated to an ERT but not an ERT itself.

Classic examples of indicator traits include ultrasonic carcass measurements and birth weight. Ultrasonic carcass measurements are a non-destructive measure of traits such as intramuscular fat percentage (IMF). Producers do not receive premiums for IMF levels, rather premiums (and discounts) are applied to quality grades. Assuming that carcass maturity values are the same, actual carcass marbling is the driver of quality grade. Although IMF is genetically correlated to carcass marbling it is not the ERT. Birth weight is another great example of an indicator trait. Selection to decrease birth weight in an attempt to reduce the prevalence of dystocia is practiced by numerous commercial bull buyers. However, birth weight does not have a direct revenue source or cost associated with it. Calving ease is the trait that has a cost associated with it. Calving ease is related to the level of assistance needed during a calving event. Although the two are related, the genetic correlation between calving ease and birth weight is between -0.6 and -0.8, suggesting that birth weight only explains 36–64% of the genetic differences between animals for calving difficulty.

Growth Traits

The earliest developed EPD for beef cattle were for birth weight (BW), weaning weight (WW), yearling weight (YW), and milk (MILK). These are still the standard EPD that are calculated for all breeds that conduct genetic evaluations.

Birth weight (BW)—Birth weight EPD reflects differences in birth weight and is used as an indicator of the probability of dystocia (calving difficulty). Birth weight is not an ERT.

Weaning weight (WW)—Weaning weight EPD predicts differences in the weight of bulls' calves at weaning. WW is an ERT for those producers who market calves at weaning.

Milk (MILK or Maternal Milk)—Milk EPD is actually maternal weaning weight, and thus reported in units of weaning weight. MILK is an ERT for producers who retain replacement females and who sell calves at weaning. In limited feed environments, selection for low to moderate Milk EPD would be warranted due to the added nutrient requirements for both lactation and maintenance.

Yearling weight (YW)—Yearling Weight EPD predicts differences in the weight of bulls' progeny at one year of age. YW is an ERT for cattle producers who might sell cattle post-weaning after a stocker program.

Dry matter intake (DMI)—Dry matter intake EPD predict differences in bulls' offspring for post-weaning feed intake. DMI is an ERT for cattle producers who retain ownership of terminal calves post-weaning.

Residual average daily gain (RADG)—This is actually an index of post-weaning gain and feed intake with changes in feed intake restricted to 0. The interpretation is differences in post-weaning gain assuming feed intake is equal. RADG is not an ERT.

Residual feed intake (RFI)—This is also an index of feed intake and post-weaning gain, but assumes changes in gain are restricted to 0. The interpretation is differences in feed intake assuming post-weaning gain is equal. RFI is not an ERT.

Total maternal (TM)—The EPD is the sum of half the weaning weight EPD and the entire milk EPD.

Yearling height (YH)—Yearling height EPD were developed as a frame size selection tool. This EPD is reported in inches of hip height at one year of age. YH is not an ERT.

Mature height (MH)—Similar to yearling height, the mature height EPD was also developed as a frame-size selection tool and is not an ERT.

Mature weight (MW)—The mature weight EPD is another indicator for maintenance energy requirements. On average, heavier cows are expected to require more feed energy in order to maintain themselves. Mature weight is an ERT given there is revenue derived from the sale of cull cows. Absent a genetic prediction for cow feed intake, it is also the best proxy or indicator trait for feed consumption of the cow herd related to maintenance.

Reproductive Traits

In addition to growth traits, breed associations have also placed an emphasis on developing EPD for reproductive traits. These traits vary from association to association and are listed below.

Scrotal circumference (SC)—Scrotal circumference is another indicator trait. The EPD for this trait is used as an indicator for the fertility of a bull's progeny through his sons' scrotal circumference and his daughters' age at puberty. The Scrotal Circumference EPD is expressed in centimeters with a large number being more desirable. SC EPD is of use only in situations in which male calves are retained as bulls. Given the availability of female fertility EPD, the utility of SC as a proxy for female fertility is diminished.

Heifer pregnancy (HP)—Heifer pregnancy is an ERT. Heifer Pregnancy EPD reports differences in the probability of bulls' daughters' ability to conceive and calve at two years of age. HP EPD is also reported as a percentage where a higher value indicates progeny with a higher probability of conceiving to calve at two years of age.

Age at first calf (AFC)—This trait is defined as the age of a female when she has her first calf. A lower value is more desirable. Differences between sires' EPD reflect differences in the average age at which their daughters will have their first calf.

Stayability (STAY)—Stayability, also called Sustained Cow Fertility (SCF), reflects the longevity of a bulls' daughter in the cow herd. This EPD predicts differences in the probability of bull's daughters having additional calves during their lifetime or remaining in the herd through extended ages.

Carcass EPD

Carcass weight (CW)—Carcass weight EPD quantifies differences in the expected carcass weight, in pounds, of a bull's progeny when they are harvested at a constant age endpoint. CW EPD is an ERT.

Ribeye area (REA)—Ribeye area EPD are reported in square inches and indicate differences in the area of longissimus muscle between the 12th and 13th ribs of bulls' offspring when slaughtered at a constant age endpoint. REA EPD is not ERT, but is a component of Yield Grade which is the ERT.

Fat thickness (FAT)—Depending on the breed association reporting the estimates, the fat thickness EPD is also sometimes referred to as the backfat EPD or just simply the fat EPD. This EPD is reported in inches and depicts differences in 12th rib fat thickness of bulls' progeny when slaughtered at a constant age endpoint. FAT EPD is not an ERT but is an indicator of yield grade which is the ERT.

Marbling (MARB)—The marbling EPD indicates differences in marbling of the ribeye of a bull's progeny when slaughtered at a constant age endpoint. Marbling is generally considered an ERT given its strong relationship to quality grade.

Yield grade (YG)—Yield Grade EPD is a prediction of differences in lean meat yield of the carcass and is an ERT given premiums and discounts are applied to YG. Phenotypically, the lower the grade, the leaner the carcass. An animal receiving a calculated yield grade of 1.0–1.9 is a Yield Grade 1, an animal receiving a calculated yield grade of 2.0–2.9 is a Yield Grade 2, etc. The highest Yield Grade is 5 so any animal receiving a calculated yield grade of 5.0 or more is classified as Yield Grade 5. Yield Grade EPD are derived using a component EPD of REA, FAT, and CW assuming a constant KPH.

Tenderness (WBS)—The tenderness EPD is reported in pounds of Warner Bratzler Shear Force such that a higher value indicates that more pounds of shear force are required to cut through the meat.

Therefore, a lower value indicates more tender meat and is more desirable. Tenderness is an ERT from an industry perspective, although producers are not currently incentivized directly for improved meat tenderness.

Management/Convenience Traits

Calving ease direct (CED)—The calving ease EPD, both direct and maternal, are the ERT. Calving ease direct EPD are a prediction of the differences of the ease at which bulls' calves will be born. Calving ease direct EPD are calculated using information from calvings of two-year-old females only (no calvings to older cows are included) and birth weight records. CED EPD is reported as a percentage so that a higher value indicates a higher probability of unassisted calving.

Calving ease maternal (CEM)—Similar to the calving ease direct EPD, the calving ease maternal EPD is also an ERT for unassisted calving. The majority of breeds, but not all, calculate CEM as total maternal calving ease (1/2 direct + maternal). Contrary to calving ease direct EPD, however, the calving ease maternal EPD predicts differences in the probability of a bull's daughters calving without assistance. CEM EPD is also expressed in terms of percentages with a higher value indicating that the bull's daughters are more likely to deliver a calf unassisted.

Pulmonary arterial pressure (PAP)—Animals with higher pulmonary arterial pressure are more susceptible to brisket (or high mountain) disease. Pulmonary arterial pressure EPD are reported in millimeters of mercury with a lower value being more desirable.

Maintenance energy (ME)—The maintenance energy EPD is a predictor of the energy needed for a cow to maintain herself. Daughters of bulls with lower maintenance energy EPD values will require less feed resources than will daughters of bulls with higher values. Therefore, it is beneficial to select bulls with lower maintenance energy EPD values. Maintenance energy EPD are measured in terms of megacalories per month.

Docility (DOC)—Docility EPD reflect predicted differences in the temperament of bulls' offspring. Animals are evaluated by producers on a scale of 1 to 6 with 1 meaning docile and 6 indicating extreme aggressive behavior. Docility EPD are reported as percentages such that animals with a higher docility EPD will have a higher probability of producing more docile animals.

Claw set (CLAW)—Claw EPD reflect differences in the claw set of offspring.

Foot angle (ANGLE)—Angle EPD reflect differences in the angle of the foot.

Teat size (TEAT)—Teat score is measured on a 1 (very large) to 9 (very small) scale and EPD are reported in units of subjective scale. Differences in sire EPDs predict the difference expected in the sires' daughters' udder characteristics.

Udder suspension (UDDR)—Udder scores are measured on a 1 (very pendulous) to 9 (very tight) scale and EPD are reported in units of the subjective scale. Differences in sire EPDs predict the difference expected in the sires' daughters' udder characteristics.

Summary

The list of available EPD continues to grow. To utilize EPD correctly, producers must develop a breeding objective to identify the traits on which they should select. Given more than one trait impacts profitability at the enterprise level, selecting on multiple traits is required. Tools to enable multiple trait selection including selection indices and decision support tools will be discussed in subsequent chapters.

References

Golden, B. L., D. J. Garrick, S. Newman, and R. M. Enns. 2000. Economically relevant traits: A framework for the next generation of EPDs. In: Proc. Of the Beef Improv. Federation Ann. Meet. & Symp., Wichita, KS. p. 2–13.

In 2022, Chuck and Judy sold their Quarter Circle U Ranch to their daughter Amy Jo and her husband, Mike Doyle, who are continuing the application of genetic science to the cattle operation.

In 2024, Chuck wrote his memoir, *My Multi-Faceted Life*, which describes his and the Backus family history.

www.ingramcontent.com/pod-product-compliance
Lightning Source LLC
Chambersburg PA
CBHW041306240426
43661CB00011B/1036